The Hot Brain

The Hot Brain

Survival, Temperature, and the Human Body

Carl V. Gisolfi and Francisco Mora

A Bradford Book
The MIT Press
Cambridge, Massachusetts
London, England

This book was set in Sabon by Crane Composition, Inc.

Printed and bound in the United States of America.

Library of Congress Cataloging-in-Publication Data

Gisolfi, Carl V.
The hot brain : survival, temperature, and the human body / Carl V. Gisolfi and Francisco Mora.
p. cm.
"A Bradford book."
Includes bibliographical references and index.
ISBN 0-262-07198-3 (hc. : alk. paper)
1. Body temperature—Regulation. 2. Brain. 3. Evolution.
I. Mora, Francisco (Mora Teruel) II. Title.
QP135.G55 2000
612.8'2—DC21 99-31037
CIP

This book is dedicated not only to twenty-five years of working together as colleagues but also to the development of a true friendship.

Contents

Preface

Aristotle was completely wrong when he said that the brain was a cold and bloodless organ, and therefore could not be the seat of thoughts and feelings. However, he was correct in his intuition that to have thoughts and feelings, a hot organ was needed. His mistake was to think that this organ was the heart. Today we know that the brain is responsible for maintaining a constant core body temperature.

Nature took almost 500 million years to achieve this physiological capacity, which evolved in mammals at a time when marked changes occurred in the size and functions of the brain. Was a constant brain temperature in part responsible for its further evolution? We do not have an answer to this question. However, with a constant core body temperature, mammals were able to change their habitats. This brought about new genetic-environmental conditions to which the brain responded, thus showing the plasticity of this organ.

Feeding, drinking, and sexual reward-based behavior are basic to survival of the individual and of the species. The control of temperature by the brain is coded along with these reward-based mechanisms. Emotion, motivation, and a sophisticated sensorimotor integration seem to form the basis of these behaviors. But they cannot occur unless the brain is functioning properly, and for it to do so, a constant brain temperature is essential. In addition to these basic survival behaviors, body and brain temperatures are implicated in many more physiological functions not restricted to the "warm bed" of "essential" functions. They are even tightly related to sleep. Could it be that at least one of the functions of sleep is to prevent the brain from becoming too hot?

Numerous questions arose in preparing this book, many of which have been raised by others in the past. For instance, why do humans regulate brain temperature at 37°C and not 20°C or 40°C? Can humans really acclimatize to the cold? Do fevers really need to be lowered, or do they have survival value? Does exercise training enable us to cope better with environmental stresses? Does it provide a measure of protection against heat stroke? Do older people lose their capacity to cope with environmental extremes, or are they just unfit? How is it possible for a 2°C rise in brain temperature during fever to cause shivering, whereas the same 2°C rise in brain temperature produced by exercise causes sweating? Can humans cool their brain below their body temperature as many other mammals can? Does this protect the brain from overheating? Is it a "hot brain" that causes us to stop exercising? The answers to some of these questions are controversial. We present the data and a viewpoint.

Included in our approach is an analysis of the strategies employed by single-celled animals, fish, reptiles, dinosaurs, mammals, and humans to cope with changes in environmental temperature. No doubt survival is at the basis of these strategies. We slowly came to the conclusion that a constant core body temperature was of fundamental importance to support higher forms of life on Earth. The interplay between behavior, body temperature, and ambient temperature could have played a crucial role in the evolution of humans. On numerous occasions throughout the book, especially in the early chapters, data are sparse and we revert to speculation. On such occasions, we have tried to make that speculation clear.

This book is written for a diverse audience. We took a broad approach to the subject of temperature regulation, but tried to maintain a focus on the brain. The information is both basic and applied, but emphasizes mechanism of action. We believe that not only students and scientists, but also nonscientists with an interest in human responses to environmental stress, the brain, evolution, and survival, will benefit from its content.

Acknowledgments

The authors wish to acknowledge the contributions of numerous colleagues who provided meaningful input to various chapters of the book. Their insights and suggestions were deeply appreciated. We thank Ana Maria Sanguinetti de Mora for her helpful suggestions and for some of the drawings included in this book. We also thank Joan Seye for typing numerous drafts of the text and for preparing the final draft of the typescript.

A very special thanks to Loring B. Rowell and Charles M. Tipton for reading virtually the entire manuscript and providing invaluable critical comments and suggestions that markedly improved the scientific quality of the book. The time and effort they devoted to this task are most deeply appreciated.

We also acknowledge the support and encouragement of Michael Rutter and MIT Press for the production of the book.

The Hot Brain

1

In the Beginning

In the Beginning Was Heat

The origin of life is subject to changing theories and controversies. However, independent of how life began, heat was always present. In fact, life, in its most reductive simplicity, had to be cooked with heat, which was the primeval source of energy (Fox and Dose 1972). The first protocells, as suggested in a recent model (Muller 1995), probably evolved around an energy-conversion unit (Granick 1957) considered to be a heat engine (Van Holde 1980). These protocells or "proteinoid microspheres" presumably were formed when polypeptides were thermally cycled as a result of convective currents in the primordial ocean (figure 1.1). Convection accounts for the initial steps in the required self-organization of the protocell. During thermal cycling, some of the proteinoids in the protocells could have acted as heat engines to condense substrates, that is, synthesize ATP by a mechanism based on a temperature-induced binding change (Muller 1995). Convection cells have always been present near volcanic hot springs and submarine hydrothermal vents. In fact, thermophilic bacteria commonly observed in these hot springs can grow and reproduce at 85°C. Thus, from the prebiotic stages on Earth, heat was a main actor on the stage where the origin of life made its debut.

With the emergence of the primitive cell came the inherent functional capacities for sensing the intensity of light and heat in the environment, and also the capacities for feeding, drinking, motility, irritability, and reproduction. Also, from the first unicellular creature on Earth, some internal mechanisms evolved that presumably included some control of the internal temperature.

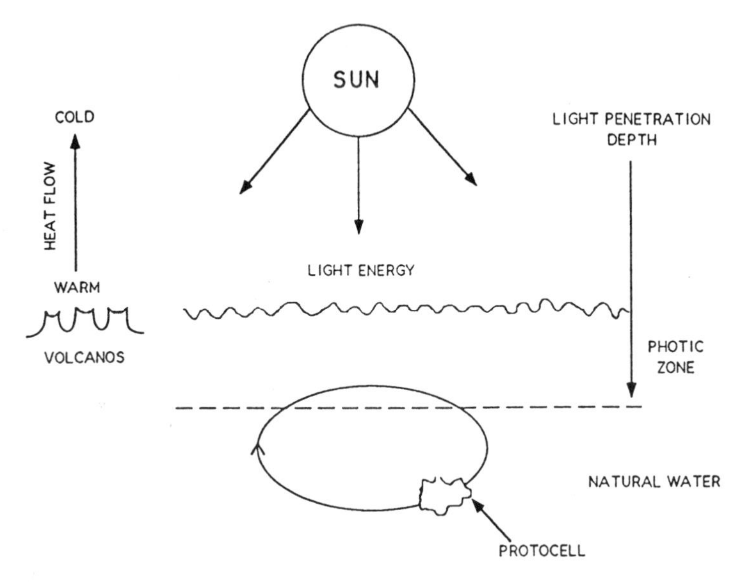

Figure 1.1
Illustration of a protocell suspended in a convecting fluid heated from below, cooled at the top, and subjected to cyclic illumination. Under these conditions, the protocell will be thermally cycled and is the best-known example of self-organization by a dissipative structure. (Modified from Muller 1995.)

It seems obvious that with such functional properties as irritability or excitability already present, unicellular organisms could react to noxious stimuli such as changes in the pH in the surrounding medium or to hot and cold environmental conditions. In fact, in primeval times, and even today, environmental thermal energy was probably one of the most important and challenging factors affecting living cells.

The Heat Intelligence of Paramecia

In an ingenious experiment, M. Mendelssohn (cited by Jenning 1906) observed that paramecia randomly dispersed themselves when ambient temperature was kept around 20°C. However, when a temperature gradient was established in their medium from 26° to 38°C or from 10° to 25°C, they congregated close to the area maintained between 24° and 28°C (figure 1.2). Moreover, when the temperature of their medium was elevated beyond their preferred temperature, these organisms became more active,

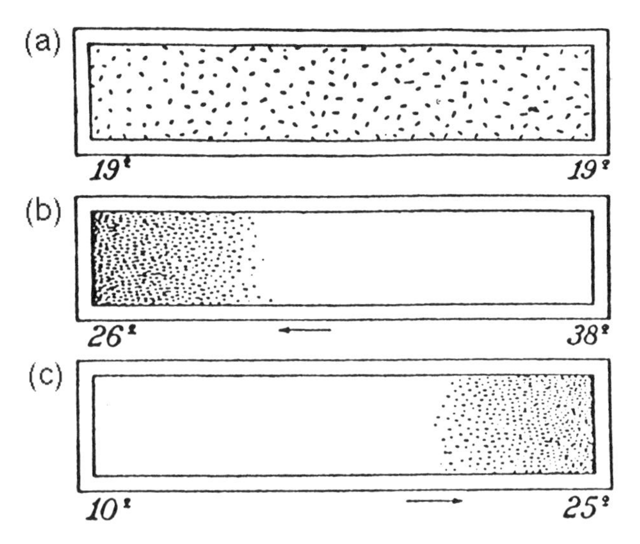

Figure 1.2
Reactions of paramecia to changes in ambient temperature. At (a) the paramecia were at a temperature of 19°C. In (b) a temperature gradient was established between 26° and 38°C. In (c) the temperature gradient was from 10° to 25°C. (Jennings 1906.)

performing a series of rotatory movements projecting them toward the cooler region. Once they reached a cooler, more comfortable environment, they resumed random locomotion. A similar, abeit slower, type of "avoiding locomotion" occurs when paramecia move from a cool to a hot environment. Jenning (1906) performed a series of experiments in paramecia that led him to conclusions similar to those of Mendelssohn. Through the microscope, Jenning observed an activation of paramecia movements as the temperature of their medium was uniformly elevated between 40° and 45°C. When a drop of cold water was placed in their environment, paramecia gathered in this region, avoiding the heated area. These experiments clearly show that unicellular organisms have the capacity to select a preferred temperature behaviorally, and therefore to protect themselves from environmental extremes.

Thus, environmental heat was a source of *deleterious stimulation,* but it was also an essential source of energy. Living organisms required heat for chemical reactions. Although some of the heat for these reactions came from metabolic activity, the main source of heat energy came from

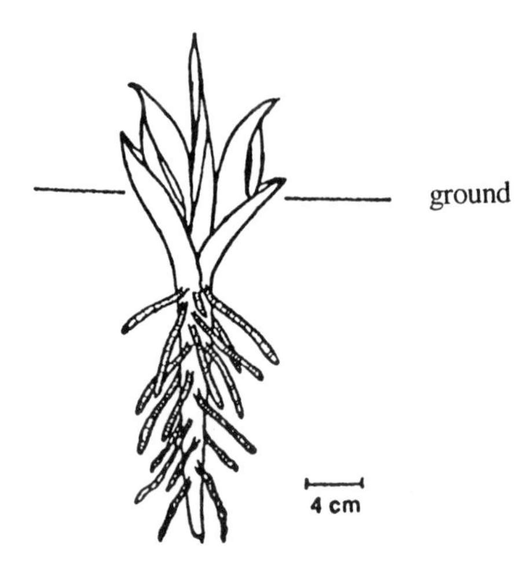

Figure 1.3
Drawing of a skunk cabbage with the spadix in the center. (From Knutson 1974.)

the environment. Therefore, the control of environmental temperature had to be coupled to the production of internal heat. It is therefore plausible to infer that unicellular organisms had some rudimentary capacity to control their internal temperature.

Temperature Control Without Nervous Tissue

When we consider multicellular organisms, we envision the coordination of all their functions by the supreme control of a nervous system. Most people would be surprised to learn that there are functions, in particular temperature regulation, that are controlled by organisms without any trace of nervous tissue. One good example is plants.

It is generally believed that the temperature of plants parallels that of their surroundings, and that plants increase their metabolic activity only during periods of elevated environmental temperature. Although this is the case for most plants, others demonstrate a self-regulation of their temperature. This phenomenon is illustrated with the eastern skunk cabbage (*Symplocarpus foetidus*) (figure 1.3) (Knutson 1974). The inflorescence (spadix; reproductive organ) of this plant (figure 1.3), weighing 2 to 9 g, maintains a temperature 15°C to 35°C above air temperature for a period of at least 2 weeks during February and March, when air temperatures are from –15°C to +15°C. This increase in temperature is maintained by an active increase in oxygen consumption and consequent heat production by the spadix (figure 1.4) (Knutson 1974).

The Protoneuron

After a long unicellular existence on Earth of more than 3 billion years, multicellular organisms (animals) appeared around 700 million years ago, and with them the innovative and powerful machinery of evolution. The adaptive advantages of multicellular organisms are many and of different kinds. Among them are cell differentiation and specialization. That is, different cells serve different functions in the same organism and, as a result, the organism is more efficient in coping with its environment. Another advantage was an increase in longevity (cells could be replaced in the multicellular organism without death of the organism). These advantages posed new problems for multicellular organisms, including the necessity for an efficient coordination among the different cell types and functions. It was under these conditions that the development of a nervous system had high selective value.

Unicellular organisms did not require a superimposed control system. They had the capacity to integrate stimuli and behave as a whole within their environment. Newly acquired functions, however, became complicated with the appearance of multicellular organisms. Specialized cells developed that were not in immediate contact with their surroundings. This created an urgent need for them to be intimately linked with other cells that were directly sensing information from the environment. This marked the beginning of the most primitive nerve cells, the protoneurons.

Most probably, in parallel with the development of multicellular organization, genesis of the nervous system took place. However, nerve cells should not be viewed as cells with completely innovative capabilities. In fact, most of the classical characteristics attributed specifically to nerve cells, such as irritability and excitability, were already present in the unicellular organism. Therefore, the origin of the nervous system must be sought in conditions present in the unicellular organism before the appearance of nerve cells.

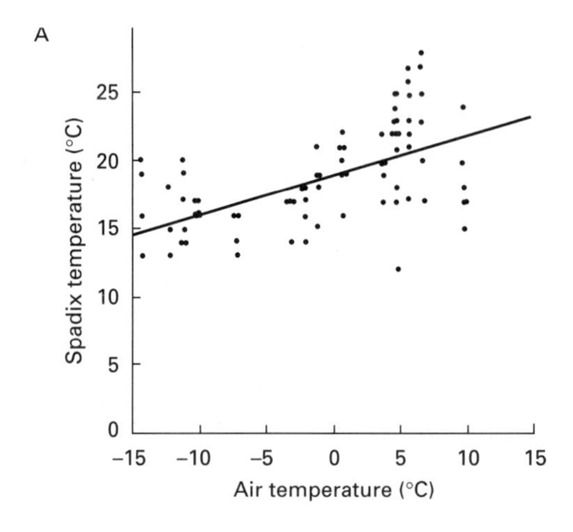

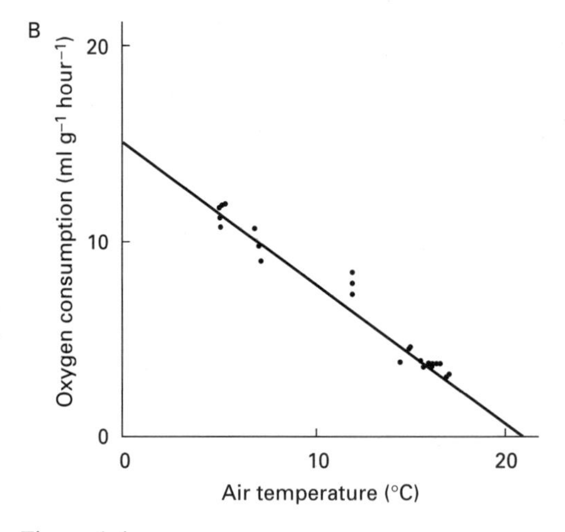

Figure 1.4
(A) Temperatures of skunk cabbage spadices at various air temperatures. (B) Oxygen uptake of skunk cabbage spadices measured in the field at available air temperatures. (From Knutson 1974.)

Multicellular Nerve Cells and Evolution

Probably one of the earliest theories about the origin of the nervous system is that of Kleinenberg (1872), who proposed that the neuromuscular cells of the hydra contained the three elementary components of the nervous system: the receptor, the conductor, and the effector. However, based on research performed on "primitive" nervous systems such as those in coelenterates, medusae, and others, several theories have been proposed to explain the origin of the nervous system. Among the most relevant are those of Parker (1919), Pantin (1956), and Passano (1963).

For Parker, "The most primitive nerve cell, from the standpoint of animal phylogeny, is the sense-cell or receptive cell, as it occurs in the sensory epithelium of the coelenterates" (1919, p. 210). Therefore, for Parker, the protoneurons initially emerged as individual sensory cells or cells for detecting changes in the immediate surroundings of the pluricellular organism. Thereafter, motor nerve cells developed.

Pantin has a more globalized conception of the primitive nervous system: "The metazoan behavior machine did not evolve cell by cell and reflex by reflex. From its origin it must have evolved the structure of the whole animal, and it must have been complex enough and organized enough to meet all the varied requirements of behavior" (1956, p. 173). In contrast with Parker, Pantin places emphasis on the simultaneous development of nondifferentiated nerve cells coordinating the whole organism in both sensory and motor responses. He further emphasizes that the primitive nervous cell was not just a single sensory cell controlling a single effector cell because a meaningful response could not be initiated by the innervation of a single muscle cell by a single receptor cell.

Passano proposed the following theory (see figure 1.5):

> Individual protomyocytes first evolved into assemblages of independent contractile cells, permitting more extensive movements than those resulting from contractions of individual myocytes. Certain of these cells became endogenous activity centers or pacemakers by developing unstable specialized membrane areas capable of active depolarization. Such local pacemakers synchronized contractions of adjacent cells by passive depolarization spread affecting the contractile mechanisms, perhaps utilizing intercellular bridges. Groups of muscle cells responding to pacemakers would permit the evolution of recurrent feeding movements.

Differentiation of these two cell types would have thus proceeded together, with what was to become nerve becoming specialized for activity initiation. Initially both would become specialized for passive conduction of depolarization. The specialization of the nerve cell for the conduction rather than the repetitive initiation of activity is seen as a secondary development in the evolution of neurons. (1963, pp. 307–308)

Passano later considered that pacemakers cells developed as sensory cells, integrating external sensory events with motor activity. In a further evolutional development of the nervous system, conducting tracks and connections (synapses) appeared, producing the first real integrated neural activity (reflexes). Further on, nerve cells concentrated into ganglia or nerve rings and sensory cells accumulated in organs associated with the ganglia.

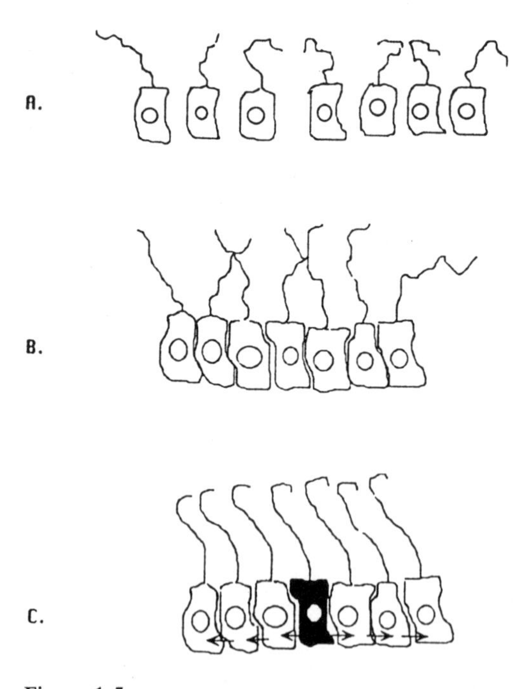

Figure 1.5
(A) Individual protomyocytes. (B) Assemblage of independent contractile myocytes. (C) It is hypothesized that after assemblage of individual protomyocytes into one single organism, one or more protomyocytes developed unstable membrane potentials (black protomyocyte) so that they could depolarize the adjacent cells.

Toward the Metazoan Organism as a Functional Unit: Microsystems of Neurons

In the development of multicellular organisms, there were protoneurons to sense changes in the environment, protoneurons to initiate coordinated movements, and protoneurons that linked these two (inter-protoneurons) (figure 1.6). These inter-protoneurons conducted information between sensory and motor cells and stimulated pacemaker activity. However, as found in some polyps today, there are certain colonial forms of multicellular organisms such as *Physalia* and *Renilla*. These organisms had different pacemakers that independently activated different parts of the primitive organism. Rather than exhibiting coordinated movements,

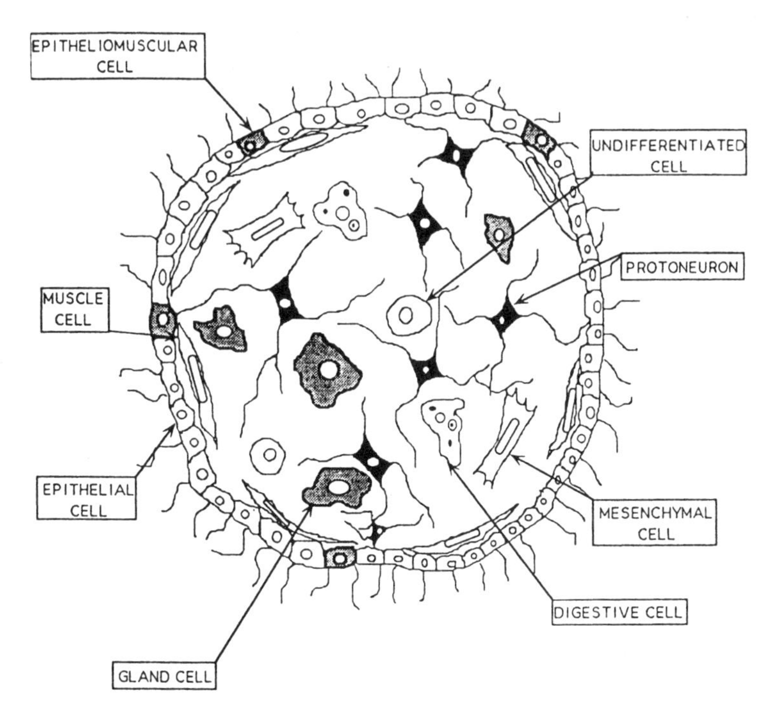

Figure 1.6
Hypothetical reconstruction of the primitive nervous system as it may have appeared in a simple metazoan. The organism may have consisted of an outer ciliated epidermis and an inner solid mesenchyme.

these primitive organisms functioned as several coexisting organisms in one body. More coordinated movements of primitive organisms probably evolved with advanced metazoans (Passano 1963).

In the same way that the protoneuron is considered an important "qualitative step forward" in the evolution of organisms, the appearance of an elementary nervous system, able to organize the complete functional activities of the metazoan as a unit, is considered a qualitative step forward. These "microsystems of neurons" (Kandel and Schwartz 1991) or "miniature nervous systems" (Jerison 1973), received the same universal acceptance as the protoneuron.

The basic system in this "miniature nervous system" is made up of a few neurons, each of which can be identified both anatomically (because of its special localization) and physiologically. The anatomic-functional relationship among these neurons is characteristic and constant for all organisms within the same species. Jerison stated:

> Their widespread occurrence reflects a fundamental evolutionary adaptation that must have occurred before the major multicellular phyla became differentiated from one another, because miniature systems have been found in many species from different phyla including mollusks, arthropods and vertebrates. At present, it seems likely that the miniature system should be regarded as the fundamental unit, or building block, with which more complex or elaborate neural systems are constructed. Its role could be as a "prewired cell assembly" in which specific nerve cells act together to carry out fairly elaborate actions. (1973, pp 9, 10)

Control of Temperature and the Metazoan Machine

One of the important activities in these primitive organisms could have been the control of temperature. The capacity to sense heat and cold, and to organize behavior to avoid extreme temperatures, was present in unicellular organisms. Moreover, multicellular organisms, devoid of nervous tissue, were able to increase their metabolism and maintain higher temperatures than their surroundings. Therefore, the capacity to control temperature was very primitive and essential for survival. Consequently, organisms that incorporated a nervous system, however primitive it was, must have incorporated the necessary circuits to control this essential function. Indeed, behavioral control of temperature through sensorimotor integration could have been a common feature of these miniature nervous systems, in both invertebrates and vertebrates. Because learning was a property of these microsystems of neurons, the control of temperature in these primitive organisms could have been flexible, and therefore became more efficient through learning and memory. In summary, the basic activities for the survival of an organism—feeding, drinking, and behavioral control of thermoregulation—could have been coded in the neural ganglia of invertebrates since ancient times. Good examples of that can be found in insects such as moths and bees (see below).

Before We Start with the Brain: Invertebrates and Temperature Regulation

Insects, which certainly possess neural ganglia with a cephalic location, but are devoid of a proper brain, regulate their body temperature in a very sophisticated fashion. Some insects thermoregulate by behavioral adjustments (ectothermic insects), but others have the capacity to generate internal heat sufficient to increase body temperature (endothermic insects) above that of the environment.

The desert locust (*Schistocerca gregaria*), some butterflies, beetles, cicadas, and artic flies thermoregulate by behavioral means. The desert locust, for example, which has a marked torpor in the morning, orients itself perpendicularly to the sun and progressively warms. This basking behavior increases the rate of heat storage, and as a consequence these locusts become more active, allowing the maintenance of a body temperature above environmental temperature.

Other insects (like bees and large moths) generate internal heat to regulate their body temperature by endothermic mechanisms (Heinrich 1974). At rest, their temperature is practically equal to that of the environment, but since a relatively high thoracic temperature is a prerequisite for flight, when they approach the period of activity, they begin shivering or warming up to increase muscle temperature. The muscles of these insects must be near or above 40°C before they can start a free flight. This expenditure of energy from muscular activity without external work (shivering) leads to the generation of large quantities of heat. In addition, these insects are covered with a dense layer of pile that prevents heat loss. These two mechanisms, increased heat production and prevention of heat

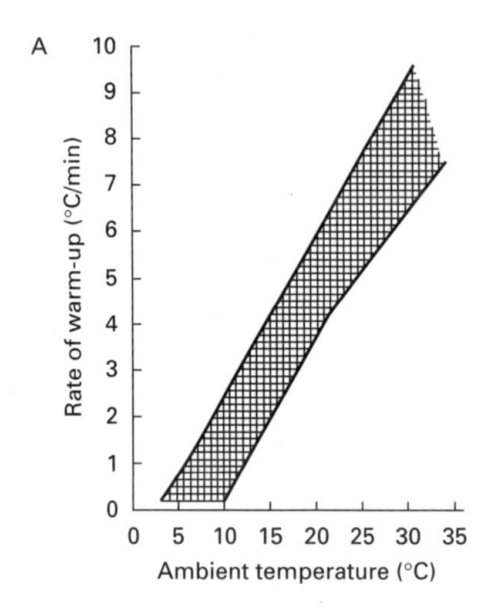

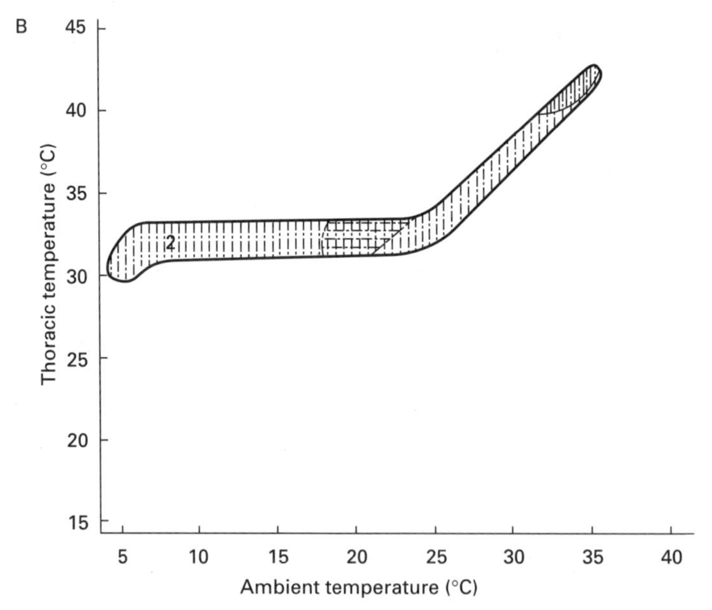

Figure 1.7
(A) Increase in thoracic temperature during shivering in moths in relation to ambient temperature. (B) Diagrammatic representation of thoracic temperatures of the bumblebee in flight in relation to ambient temperature. (From Heinrich 1974.)

loss, allow them to increase their thoracic temperature by more than 20 to 30°C above ambient temperature. Their capacity to retain this high thoracic temperature during flight is excellent. In fact, it has been shown that moths and bumblebees may fly with a thoracic temperature of 36°C at an ambient temperature of 3°C (figure 1.7). It is also of interest that colonies or clusters of honeybees (of some 50 or more individuals) behave as a unique organism, regulating temperature of the core of the cluster (in both winter and summer) at about 34° to 35°C. This activity must be successfully coordinated, because bees will die if they remain 1 or 2 days at a temperature as low as 8°C (Folk 1974).

Unfortunately, very little is known about the neurophysiology of thermoregulation in insects. There are neurophysiological experiments showing that exteroreceptors provide sensory input to reflex mechanisms that are used to maintain specific high speed, and also that a series of reflexes maintains the flight altitude of the insects; secondarily, these reflexes keep flight muscles working to maintain them above some minimum temperature. In locusts and moths it has been shown that contraction of muscles (and the corresponding heat production) is produced by changing the number of motor units recruited and by activating them multiply rather to achieve a stronger muscle contraction (Heinrich 1974).

2

Evolution and the Control of Body Temperature as a Mechanism for Survival

Evolutionary Perspectives

Through evolution from unicellular organisms to water-land vertebrates, animals had to avoid extreme, lethal environmental temperatures to survive. Presumably, animals acquired, through natural selection, many different advantages to cope with different thermal environmental stresses. These advantages, useful for the survival of the species, should have been stored in the brain and passed from generation to generation. Therefore, one would expect that studying the brain of living representatives of primitive animals and of modern species would provide some insight into how different thermal strategies evolved, and perhaps even into the evolution of homeothermy. Unfortunately, very little is known, even in living species, about the evolution of the neural substrates involved in the control of body temperature. Thus, all inferences about the evolution of the brain in relation to body temperature control will be primarily descriptive and speculative.

During evolution, two main events regarding temperature can be envisaged that could have had important implications for the survival of animals. The first was the movement of animals from a relatively constant thermal environment provided by water to wide variations in environmental temperature on land (crossopterygians and tetrapods). For them to survive, there must have been changes in the body, as well as in the brain, to sense temperature fluctuations in this new environment (land) and to behave accordingly. Therefore, it makes sense to speculate that during the transition from water to land, important changes must have occurred in the programming of the brain to control body temperature by behavioral means. More than that, this dramatic event put tremendous pressure on these animals to develop different behavioral strategies (implicating brain and body) to adapt to constantly changing environmental conditions. For instance, the fins used for swimming were now adapted to walking on land. Would this not require important changes in the motor programming of the brain? And what about changes in brain programming of behavioral temperature regulation?

The second important event in the evolution of thermoregulation was the development of homeothermy, which seems to have occurred in primitive nocturnal animals. Under conditions of poor solar radiation, these animals lost their capacity to gain heat from the environment; as a consequence, one could envisage them behaving torpidly. Over time, these animals developed a remarkable capacity to produce significant amounts of heat internally. These changes must have been produced by changes in the brain as well as in the body to allow them not only to cope with the new sensorial situation (absence of light) but also to have autonomic control of body temperature. How could this have occurred?

Before entering into these general considerations, however, in order to have a better understanding of a phylogeny of thermoregulation, let's trace the geochronology and geobiology of vertebrates since they began to appear, most probably in the middle to late Ordovician period, about 500 million years ago. Today, there are approximately 46,000 living species vertebrates and 1,250,000 species of invertebrates. Table 2.1 shows the biogeological time scale.

The Long Path from Cold-Blooded to Warm-Blooded Animals

Outside the area of physiology, the terms "cold-blooded animals" and "warm-blooded animals" involve two rather confusing concepts. Do cold-blooded animals really have cold blood? And if so, what about the rest of the body? The same questions could be posed for warm-blooded animals. Because these are vague concepts used throughout the literature of thermoregulation, for the sake of clarity they will be described here before discussing evolution.

Cold-blooded animals (invertebrates, fish, frogs, and reptiles) are poor heat generators. That is, they produce internal heat, but not in quantities

Table 2.1
Geologic periods after the time when fossils first became abundant

Era (and duration)	Period	Estimated time since beginning of each period (in millions of years)	Epoch	Life
Cenozoic (age mammals; about 65 million years)	Quarter-nary	2+	Holocene (recent)	Modern species and sub-species: dominance of man.
			Pleistocene	Modern species of mam-mals or their forerun-ners; decimation of large mammals; wide spread glaciation.
			Pliocene	Appearance of many modern genera of mam mals.
			Miocene	Rise of modern sub-families of mammals; spread of grassy plains; evolution of grazing mammals.
	Tertiary	65	Oligocene	Rise of modern familie: of mammals.
			Eocene	Rise of modern orders and suborders of mam mals.
			Paleocene	Dominance of archiac mammals.
Mesozoic (age of reptiles; lasted about 165 million years)	Creta-ceous	130		Dominance of angio-sperm plants commences; extinction of larger rep-tiles and ammonities by end of period.
	Jurassic	180		Reptiles dominant on land, sea, and in air; first birds; archaic mammals.

Table 2.1 (continued)

Era (and duration)	Period	Estimated time since beginning of each period (in millions of years)	Epoch	Life
	Triassic	230		First dinosaurs, turtles, ichthyosaurs, plesiosaurs; cycads and conifers dominant.
Paleozoic (lasted about 340 million-years)	Permian	280		Radiation of reptiles, which displace amphibians as dominant group: widespread glacition.
	Carboniferous	350		Fern and seed fern coal forests; sharks and crinoids abundant; radiation of amphib ians; first reptiles.
	Devonian	400		Age of fishes (mostly fresh water); first trees, forests and amphibians.
	Silurian	450		Invasion of the land by plants and arthropods; archaic fishes.
	Ordovician	500		Appearance of vertebrates (ostracoderms); brachipods and cephalopods dominant.
	Cambrian	570		Appearance of all major invertebrate phyla and many classes; dominance of trilobites and brachiopods; diversified algae.

From Romer, 1970.

sufficient to maintain a constant body temperature. Therefore, their body temperature fluctuates according to changes in the temperature of their surroundings (well beyond 2°C; see figure 2.1). To elevate their body temperature sufficiently to become active, these animals depend upon behavioral thermoregulation, which results in the accumulation of heat from the sun (a condition also called *heliothermy*). These animals are also called *ectotherms,* from "ecto" (outside) and "therm" (temperature)—regulating body temperature from an external source of heat. Two synonymous terms used interchangeably with "ectotherms" are *poikilotherms* and *heterotherms,* from the Greek *poikilos* (variable) and *heteros* (different). Poikilotherms have often been called *conformers,* in contrast to *regulators* (homeotherms).

It is generally agreed that the so-called warm-blooded animals (birds and mammals) maintain a high, constant body temperature that depends upon their own production of internal heat. A high rate of metabolism is the source of this internal heat. Warm-blooded animals are also called *endotherms*, from "endo" and "thermal" (internal and temperature). This refers to the fact that heat is produced internally. Because of a high metabolic heat production, the thermal state of the endothermic animal is such that body temperature is maintained almost constant over a wide range of environmental temperatures and despite effective heat-loss mechanisms. Animals in this category are also called *homeotherms,* from the Greek *homos* (fairly equal) and *thermos* (temperature), and *regulators.*

However, as mentioned in chapter 1, some invertebrates developed different strategies to allow them to thermoregulate as if they were endotherms. And there are vertebrates that thermoregulate like ectotherms. For instance, a kind of homeothermy and even real endothermy has been achieved in fish such as the tuna. Also, in reptiles such as dinosaurs, a kind of homeothermy has been proposed because of their enormous mass/surface area ratio. This last condition has been called *gigantothermy, ectothermic homeothermy, inertial homeothermy,* or *mass homeothermy*. And mammals (at least primitive mammals), considered without exception to be homeotherms, have exceptions among them. For instance, marsupials and monotremes are considered not to have a constant body temperature although they are not hibernators. Some mam-

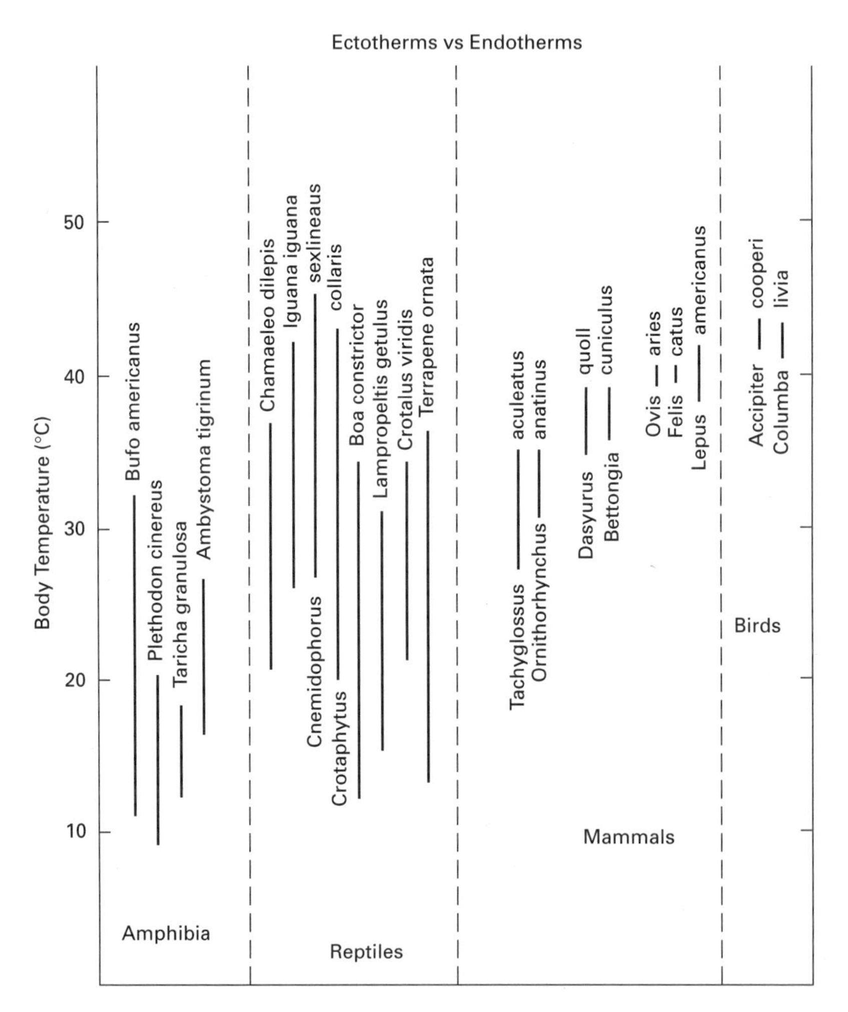

Figure 2.1
Variations in the range of active body temperature among ectotherms and endotherms. Some ectotherms (reptiles) experience temperatures as high as, or higher than, those of some endotherms. (Modified from Ostrom 1980.)

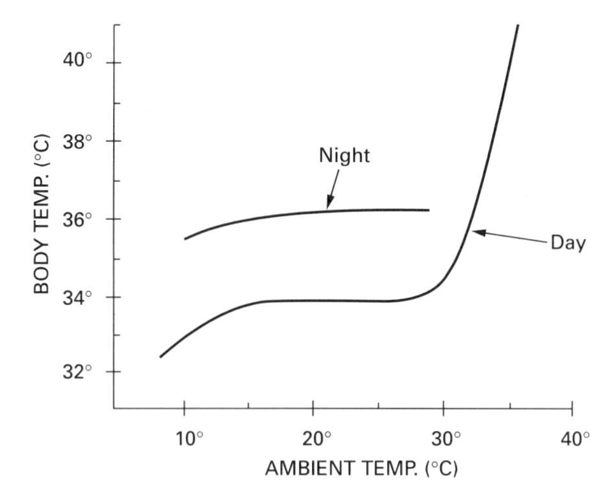

Figure 2.2
Body temperature as a function of ambient temperature in an active Central American opossum, *Metachirus,* during day and night. (Modified from Morrison and Ryser 1952.)

mals have dormancy or hibernating periods during which their body temperature decreases markedly in accordance with the temperature in their surroundings (figure 2.2).

It looks as if interactions between the physical characteristics of the environment and anatomical and physiological characteristics of some animal species are powerful determinants of the capacity of animals to adapt to changes in environmental temperatures. Among these characteristics, apart from temperature itself, are the intensity and duration of solar radiation, wind velocity, relative humidity, size and shape of the animal, quality and quantity of insulation, and metabolic rate.

These considerations support the view that different selective pressures led animals to develop different thermal strategies to adapt to particular thermal niches or climatic spaces. Spotila (1980 p. 251) stated, "We can best visualize pathways in the development of homeothermy not as proceeding sequentially from one evolutionary level to the next, but rather as divergent courses, in which a common problem has been resolved in different ways." (See figure 2.3.)

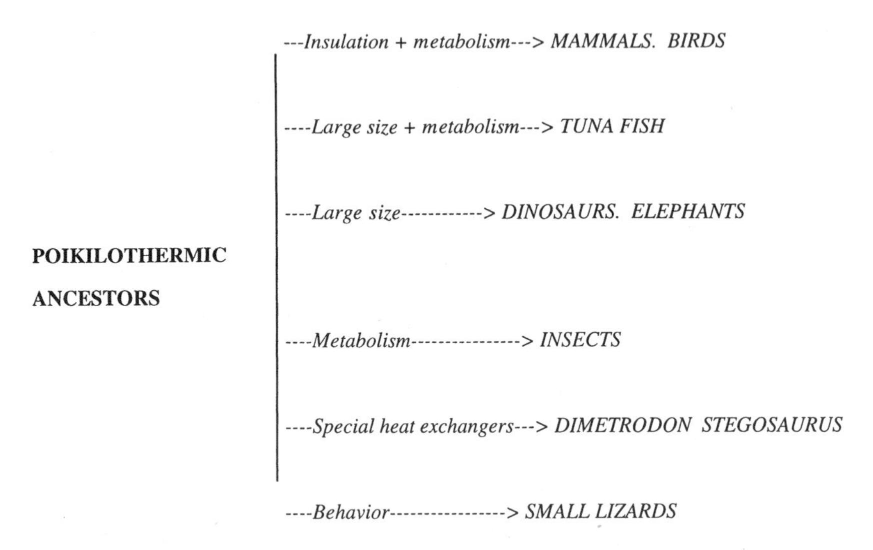

Figure 2.3
Evolutionary strategies employed by poikilothermic ancestors in their transition to homeothermic and poikilothermic present-day animals to increase and/or maintain body temperature. (Modified from Spotila 1980.)

Fish, Frogs, and Crocodiles: An Evolutionary Reconstruction

Figure 2.4 shows the main evolutionary trends leading to modern mammals. Six main steps could be envisaged in this general trend:

1. Ancestral protochordates gave rise to the *Agnathans* (ostracoderms), or fish without jaws. These archaic fish first appeared in the Ordovician period of the *Paleozoic era*, around 500 million years ago. There are fossil records of these fish. They also have present-day representatives, the (*cyclostomes*). Examples include hagfish and lamprey.
2. The Agnatha gave rise to the *placoderms*, jawed fish. Living representatives of placoderms are the *elasmobranchs* (sharks and skates) and the *holosteans* (sturgeon, gar) and *teleosts* (perch).
3. *Chondrichthyes* and *Osteichthyes* evolved from the *placoderms*. The Chondrichthyes are cartilaginous fish. The Osteichthyes are bony fish. The latter class gave rise to the *actinopterygians* and *sarcopterygians*. Descended from *sarcopterygians* are the *crossopterygians* (from which land animals appear to have descended). These have a living descendent, the coelacanths (*Latimeria*). The dipnoans (*lungfish*), also descended from the sarcopterygians, have living descendants in the tropical regions of Australia, Africa, and South America.

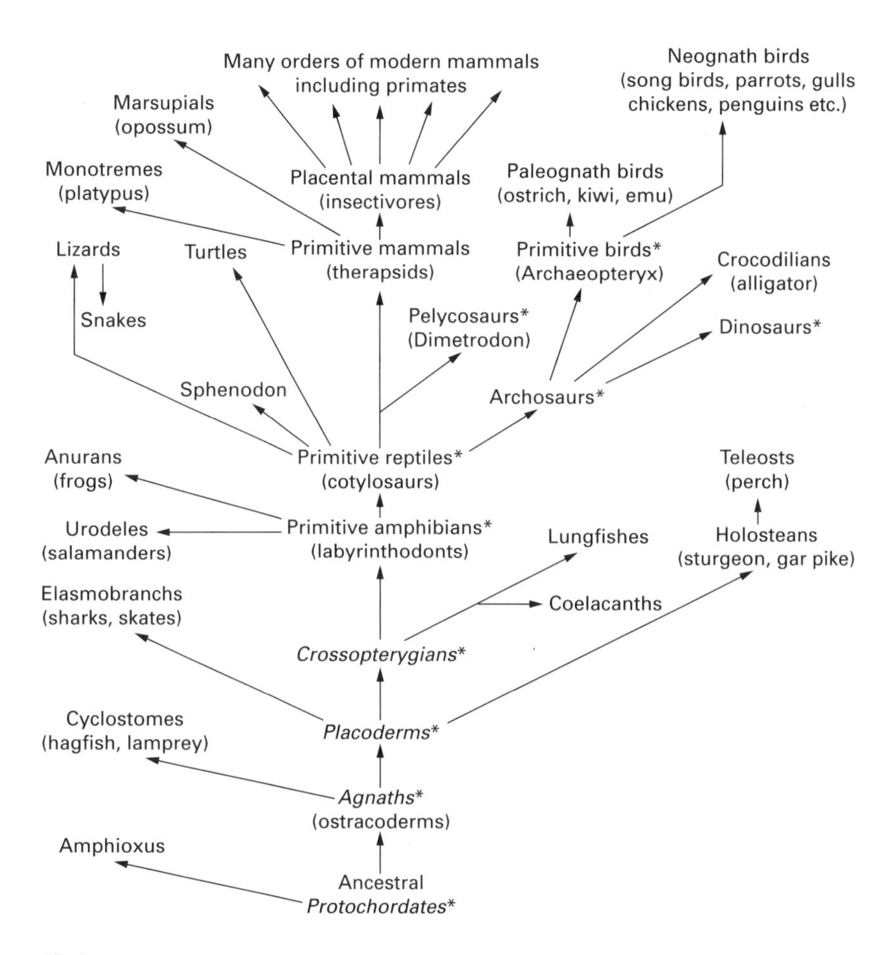

Figure 2.4
Schematic diagram illustrating vertebrate evolution. (From Romer 1970.)

4. The descendants of the crossopterygians are the primitive amphibians that appeared at the end of the Devonian period, around 400 million years ago. Descended from these primitive amphibians are the *urodeles* (salamanders) and *anurans* (frogs).
5. From the *amphibians* the *reptiles* arose in the middle of the Carboniferous period, around 350 million years ago.
6. The *primitive reptiles* gave rise to the *therapsids primitive mammals. Archosaurs* also evolved from the primitive reptiles and gave rise to *primitive birds*.

Is there any evidence that during the Paleozoic (340 million years ago) and Mesozic (165 million years ago) eras, up to the appearance of primitive mammals, animals developed any thermoregulatory strategies different from those existing in their living descendants or living primitive representatives? The answer is no. Nonetheless, there are several points worth considering.

Present-day fish, frogs, and reptiles are cold-blooded animals. However, that doesn't mean they were unable to maintain a body temperature higher than their environment. In general, when the environment changes rather permanently, this elicits adaptive responses that can be structural (long-term adaptations) or regulatory adjustments (temporary adaptations). In these animals, long-term adaptations entailed species-specific genetic adaptations. For instance, certain Antarctic fish inhabiting ice-laden seas swam without freezing in waters that had a temperature of –1.9°C. This is so because these fish had the capacity for synthesizing a unique "antifreeze" (glycoprotein) (De Vries and Lin 1977).

Other adaptations may have involved biochemical and/or endocrine changes, but not structural alterations. For instance, the concentration of cytochrome C (a component of the mitochondrial respiratory chain) increases when green sunfish (*Lepomis cynauellus*) are moved from 25°C to 5°C. In these fish, cold produces a 40 percent decrease in cytochrome C synthesis, but a 60 percent decrease in the rate of degradation. The resulting increase in cytochrome C, together with other possible changes in the mitochondria, could account for the rise in heat production through a higher metabolic rate, and therefore acclimation of these fish to the new, cold environment. Thyroxine not only can alter the behavioral response of fish to temperature changes but also can influence their general level of spontaneous activity, chill resistance, and ability to sense salinity (see figure 2.9) (Stevens 1973).

Now, let us look at our primitive ancestors. As shown in figure 2.4, the first vertebrates, the ostracoderms (Agnatha), appeared around 500 million years ago. These animals, most probably living in freshwater, would have developed behavioral skills to avoid changes in temperature beyond those to which they were acclimated. In fact, all mobile organisms studied to date, including unicellular organisms without a central nervous system, have the ability to avoid adverse environmental temperatures.

One approach to understanding the thermoregulatory capacities of these ancient fish would be to study the responses to different thermal challenges of their living descendants, lamprey and hagfish. Unfortunately, very few investigations have studied these fish. Crawshaw and colleagues (Crawshaw, Moffitt, et al. 1981) attempted to investigate the thermoregulatory behavior of a lamprey, *Lampetra tridentata*, but their attempts met with frustration. They were able, however, to demonstrate that lamprey prefer waters of about 16°C, the usual temperature of the rivers they ascend in the spring. They also observed that when they changed the temperature of the water in which the lamprey were immersed, the breathing rate and heart rate changed much more rapidly than mean body temperature. These results were the same as those reported by the same authors in carp (cyclostomes). It seems that lamprey react to temperature changes in their environment as any other fish does, by activating, in reflex fashion, some functions of the vegetative nervous system (Lemons and Crawshaw 1978).

From these observations we conclude that very little can be said about the thermoregulatory behavior of these fish, apart from the inference of body and behavioral reactions to avoiding damaging external temperatures. Crawshaw (Crawshaw, Moffitt, et al. 1981) concludes that since the ostracoderms became extinct about 400 million years ago, it is difficult to assess how they dealt with varied thermal environments. All vertebrates alive today, however primitive their structure, have had a long time to evolve specialized physiological capabilities. However, if we investigate many living vertebrates from varied classes and inhabiting various environments, and we find that they all exhibit certain characteristics, we can conclude that ancestral vertebrates had developed those characteristics or at least embodied the predisposition for such development.

These statements are reinforced by the similarity in anatomical characteristics between the fossil vertebrates of about 500 million years ago—armored ostracoderms, jawless fish (Agnatha) and their relatives—and modern lamprey and hagfish (Romer 1970).

The jawless fish (500 millions years ago) had a brain that could be considered a model of the vertebrate brain. Figure 2.5 shows the brain of a jawless primitive fish (*Heterostraci*) and a living jawless fish (lamprey).

The main anatomical divisions of the brain are already present in jawless fish. Unfortunately, the neuroanatomy of the brain of Agnatha has not been well studied, and little is known about key areas chiefly responsible for temperature regulation (the preoptic area and the hypothalamus). In recent neurohistological studies performed on the brain of the lamprey, a preoptic area and a dorsal-ventral division of the hypothalamus have been recognized.

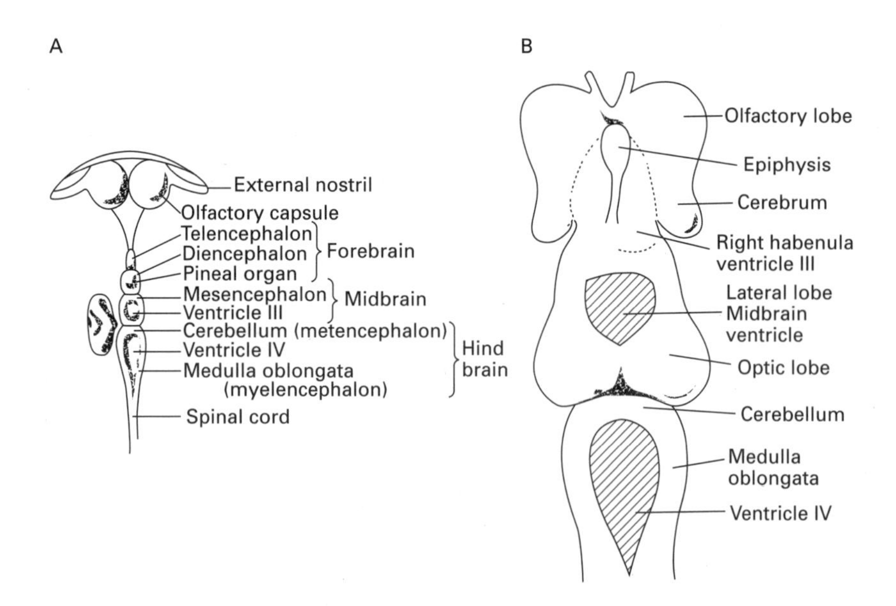

Figure 2.5
(A) Reconstruction of the brain of an extinct jawless fish (*Heterostraci*). (B) Dorsal view of the brain of a lamprey (*Petromyzon*). (A from Jerison 1973; B from Romer 1970.)

Bony Fish

Most of the research on temperature regulation in fish has been done in bony fish (see evolutionary tree in figure 2.6). These fish are sensitive to changes in water temperature of less than 1°C. Crawshaw and Hammel (1971) have shown in a series of elegant experiments on Antarctic ice fish that when these animals are placed in a two-chambered tank with one chamber at 3°C (the temperature at which they acclimated) and the other at 5°C, they behave quite characteristically. When placed in the 5°C chamber, they escape rapidly to the 3°C chamber. This behavior argues in favor of a kind of reflex arc in which the input is quickly reflected in the output, and reinforces the idea of neural reflexes as the basis for temperature regulation in fish. Also, under sudden changes in external environmental temperature, increases in breathing and heart rates and other autonomic reactions occur before changes in mean body temperature are detected. This should not be surprising because many autonomic functions (cardiovascular, respiratory, energetic, ionic, and osmotic responses) seem to share the same areas of the brain in fish. In fact, Crawshaw and Hammel (1974) have shown that heating/cooling the anterior portion of the brain of fish provokes swimming to cooler/warmer water, respectively.

Nelson and Prosser (1979) demonstrated that after destruction of the anterior part of the brain stem, fish are unable to select a particular tem-

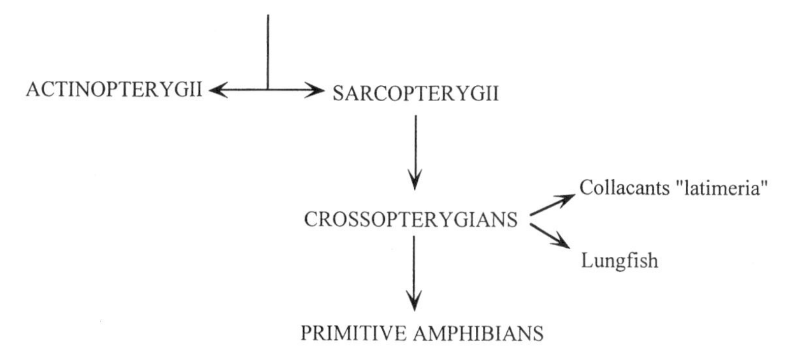

Figure 2.6
Simplified evolutionary tree of bony fish. (Modified from Romer 1970.)

perature when they are placed in a thermal gradient. Hammel et al. (1973) have described a simple model in which a "reflex circuitry" located in the anterior part of the brain stem of fish could subserve these functions (figure 2.7). A basic neuronal wiring of this type located in the brain stem and subserving reflex activity could have served the survival of fish for millions or years. Nonetheless, fish have a certain degree of internal activity to select ambient temperatures depending on changes in their internal milieu. For example, during winter starvation, certain fish select temperatures lower than those normally selected. This strategy could include hormones that interact with the brain. Under these conditions, fish have a reduced metabolic rate and therefore require less food. An obvious consequence of this is an increase in survival possibilities.

Crossopterygians

Land vertebrates seem to have descended from the crossopterygians. In the Devonian period, 400 million years ago, these ferocious and predaceous fish were the most common bony fish (figure 2.8). They showed structural features that made them suitable for life on land. They had lungs (as did some of their ancestors the placoderms) and sturdy fins that could be used as limbs. It is from this order of fish that amphibians evolved. In the Carboniferous period, 350 million years ago, these fish, which included the *rhipidistians*, started to disappear, and they became extinct at the end of the Paleozoic era.

There is evidence that in the Devonian period there were dry periods during which freshwater pools decreased in size, and consequently many fish became extinct. Under these conditions, the survival of fish with lungs and strong fins was substantial. Fish with these characteristics could breathe and move over land from one pool of water to another. Moreover, the rhipidistians were carnivores, living upon other fish, and in the drying pools they presumably found a considerable number of fish, alive or moribund, that could supply them with food.

From these observations, the conquest of land was really accidental, because nothing was on land for a carnivorous fish to eat until insects became widespread in the Carboniferous period. Therefore, these fish must have lived on other fish for a long time before they evolved into amphib-

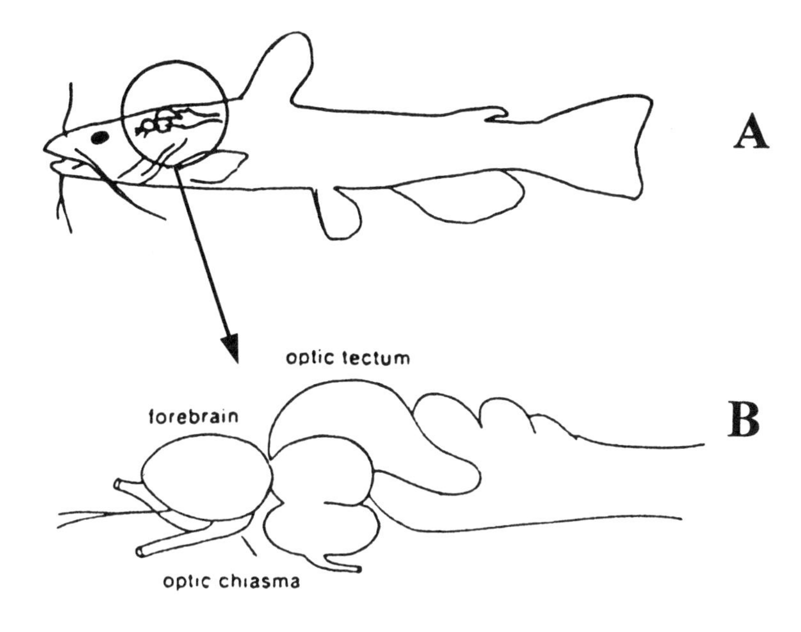

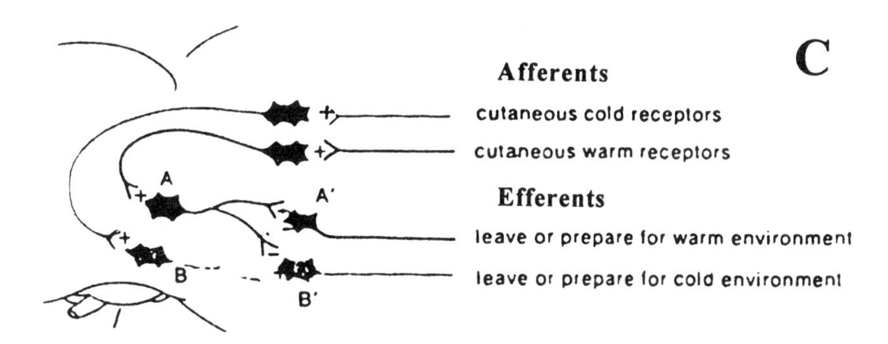

Figure 2.7
(A) Schematic diagram of the fish brain. (B) Anterior portion of the brainstem, enlarged. (C) Simplified neuronal network illustrating a negative feedback loop. Stimulation of cutaneous warm or cold receptors activates a behavioral response to counter the deviation in environmental temperature. (Modified from Crawshaw et al. 1981.)

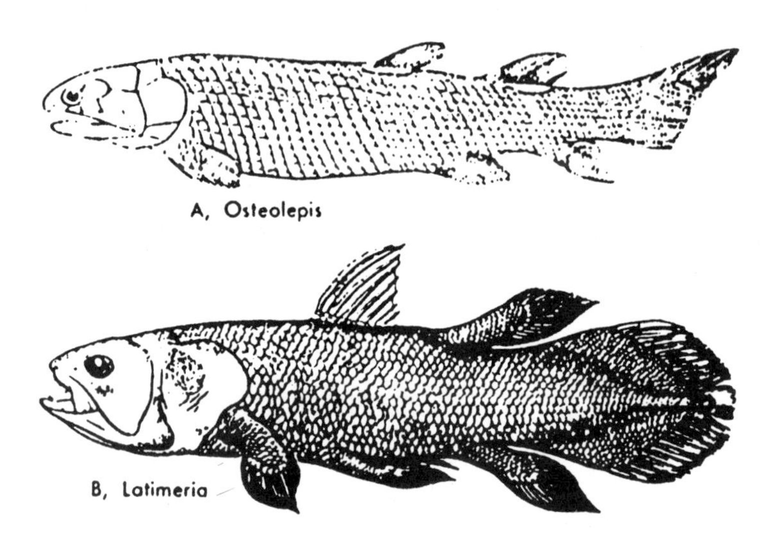

Figure 2.8
Examples of crossopterygians. (A) Devonian form. (B) Living coelacanth (*Latimeria*). (Taken from Romer 1970.)

ians. What occurred within the body and brain of crossopterygians during the period of their invasion of land that lasted 20 or 30 million years, with regard to the control of body temperature?

As we have seen in preceding sections of this chapter, water-breathing fish had a body temperature close to the temperature of the water they inhabited. However, a dramatic change occurred when they invaded land, because the loss and gain of heat increased markedly. On land, animals were much more dependent upon changes in ambient temperature. We speculate that gradual changes occurred in the brain and skin of these animals (inherited by their descendants, amphibians and reptiles) that allowed them to sense the intensity of solar radiation and to develop behavioral skills to move in and out of the sun. Parallel to that is the importance for these primitive land invaders to avoid dehydration, which probably had greater consequences than changes in environmental temperature.

These last speculations, however, cannot be substantiated by any special findings in the fossil records or in the living descendants of these fish, the "living fossil" *Latimeria*. Romer (1937) described the endocranial cavity of a rhipidistian of about 300 million years ago and concluded that this fish (*Ectosteorhachis nitidus*) had no special adaptations in its brain different from the general pattern found in the lower vertebrates as a group. In fact, not until the emergence of birds and mammals did the brains show a tendency toward the enlargement of superficially visible structures (Jerison 1973).

Latimeria seems to have a generalized type of "early vertebrate brain" like that of *E. nitidus*. However, its internal anatomy says very little about their ancestral predecessors, Jerison (1973).

In the living descendant of crossopterygians, lungfish, the preoptic area and hypothalamus are present. However, detailed histological studies have not been performed in this class of fish. Most descriptions of this fish are based on extrapolations made from frogs and salamanders.

In conclusion, it looks as if the crossopterygians, from which the land vertebrates evolved, had brains that followed a generalized and primitive vertebrate pattern described by Jerison (1973). This is in contrast to the brain of the actinopterygians, at present the most successful fish in which special adaptations of the brain occurred.

Frogs and Crocodiles

As Romer (1970 p. 60) pointed out, "Greatest, perhaps, of all ventures made by the vertebrates during their long history was the development of tetrapods in the invasion of the land" (figure 2.9). The derivation of four-legged walking animals from fish resulted in structural changes in every anatomical and physiological system of the body. In contrast, the brain of both amphibians and reptiles did not change in any meaningful way from the generalized and primitive vertebrate pattern of their predecessors, the crossopterygians. In fact, present studies of frogs, crocodiles, and turtles suggest that they have undergone very small changes with respect to their predecessors in the Triassic period (Jerison 1973).

Myhre and Hammmel (1969) stressed that in reptiles, as in fish and frogs, the anterior brain stem plays an important role in temperature regulation. Frogs and reptiles, like fish, are poikilothermic. Their internal temperature fluctuates several degrees centigrade daily and from season to season. They do not produce enough internal heat, nor do they have ef-

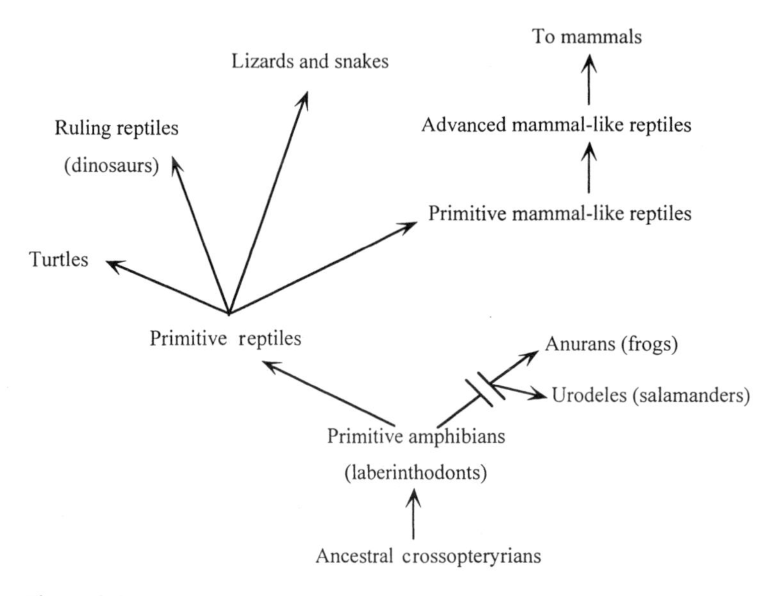

Figure 2.9
Family trees of amphibians and reptiles from the ancestral crossopterygians. (Modified from Romer 1970.)

fective insulation. However, through behavior (alternately basking and seeking shade), they can control and maintain body temperature above that of their surroundings (Greenberg 1980). Therefore, the activity of these animals is tightly coupled to energy obtained from the sun. Strategies adapted by different species to solve the problem of maintaining a certain constancy of body temperature range from behavioral adjustments (the most important) to rudimentary vegetative responses (in extreme situations) such as panting (turtles and lizards) or shivering (pythons). Leatherback turtles are an extreme case of successful adaptation to the cold by a reptile. These turtles can maintain a constant body temperature above 25°C at water temperatures below 7°C. Paladino and Spotila (1994 p. 261) explain that "This dramatic ability to maintain warm temperatures in cold, highly conductive water, that would quickly cause hypothermia and kill most of the endotherms, is made possible by a mechanism we describe as gigantothermy which is the ability to maintain constant warm body temperatures with low energy consumption, control of peripheral circulation, and extensive insulation due to large body size."

In summary, during the evolution of lower vertebrates, behavioral thermoregulation seems to have been chosen by nature as the primary strategy to cope with changes in environmental temperature. However, as observed in lizards and turtles, some rudiments of autonomic thermal responses to extreme environmental temperatures also were coded in the brain of these lower vertebrates.

Dinosaurs: A Special Story in Thermoregulation

Dinosaurs lived on Earth for more than 150 million years. Compared with the existence of other species, including humans (with only about 2 million years), they were a extremely successful experiment of nature. Unfortunately, because dinosaurs have been extinct for more than 65 million years, there is no way of determining the real lifestyle or physiology of these large reptiles. Regarding temperature regulation, only through inferences obtained from fossilized bones, the anatomy of complete skeletons, reconstructions of the brain, and comparative studies with present living reptiles can we obtain any clues about the many fascinating aspects of these creatures.

The first obvious impression one receives about the physiology of dinosaurs is that they cannot be treated as if they were just "reptiles" or a single group of homogeneous animals (although all of them belong to the class Reptilia). From studies of more than 2,000 complete fossilized skeletons (Sheehan 1994), dinosaurs have been classified in more than 27 families with different anatomies, bone histology, postures, and gaits, and supposedly different hemodynamics, activity levels, feeding habits, and presumed predator-prey relationships (Ostrom 1980). Therefore, possibilities for dinosaurs having developed different physiological strategies for solving the problem of temperature regulation during a period as long as 165 million years (the Mesozoic era) are not lacking. Knowing how dinosaurs regulated body temperature may help us understand how they lived and behaved, and also provide insight into their diversification, and even their extinction.

Since the mid-1970s, a considerable literature has accumulated that provides different indirect arguments for and against dinosaurs being ectothermic or endothermic (Thomas and Olson 1980). Although most of the literature supports the idea that dinosaurs were ectothermic, Robert Bakker (1986) pioneered the idea that dinosaurs, in particular the large Cretaceous dinosaurs, were endotherms. Paladino and Spotila (1994, pp. 263–264) summarized the factors supporting endothermy as follows:

1. Fossil evidence indicates that late Cretaceous dinosaurs lived in the polar more seasonal areas where there might have even been snow. Dinosaurs had to be endotherms to accomplish this;
2. Oxygen 18 isotope analysis for the bones of dinosaurs indicates they maintained high, constant body temperatures even in their extremities, like modern birds;
3. Dinosaurs had complex social behaviors such as herding, communal nesting, and long distance migrations which required them to be endothermic;
4. Dinosaurs appear to have richly vascularized, dense haversian bone with growth plates similar to those of birds;
5. Birds evolved from small bipedal dinosaurs that diverged from the dinosaurian lineage more than 90 million years ago; thus all dinosaurs (a monophyletic group) must also have been endotherms since they are the direct precursors of birds;
6. Since dinosaurs were very active, they needed an increased aerobic capacity that necessitated the development of endothermy [the increased aerobic capacity hypothesis for the evolution of endothermy].

The truth probably lies somewhere between endothermy and ectothermy. Research in living species strongly supports the possibility of dinosaurs having been ectothermic homeotherms, a condition also called gigantothermy. This condition is reached by animals that gain most of their body heat from the environment (ectotherms) but, because of their mass/surface area ratio, maintain a constant high body temperature (homeotherms) (Spotila 1980). This possibility has received strong theoretical and experimental support from comparative studies on large living reptiles (Paladino and Spotila 1994).

To complicate matters, Barrick (1994) suggests that large dinosaurs were (1) endotherms throughout their life; (2) endotherms as juveniles and mass homeotherms as adults; or (3) ectotherms as juveniles and mass homeotherms as adults. Added to this is the recent suggestion, based on isotope composition of bone, that because dinosaurs were relatively fast-paced, grew rapidly, and probably had a high metabolism, most of them were endotherms. In objectively evaluating the evidence above, McGowan (1991, pp. 162–163) draws the following conclusions :

1. Most dinosaurs, because of their large size, could not have avoided maintaining a fairly constant body temperature, a strategy described as inertial homeothermy. It may safely be assumed that hadrosaur-sized dinosaurs, and larger ones, were inertial homeotherms.
2. The body temperature of the inertial homeotherms may have been relatively high, comparable to those of modern birds and mammals. Daily fluctuations may have been only a degree or two, but seasonal variations may have been greater.
3. The metabolic rate of an inertial homeotherm may have been a little higher than that of a modern reptile of similar weight, but it would probably have been more typical of modern reptiles than of birds and mammals.
4. Erect posture would have given dinosaurs a more efficient respiratory system than that of modern lizards. Together with other modifications, including a four-chambered heart, improved respiration would probably have given them a fairly wide aerobic scope, permitting them higher levels of aerobic exercise than modern reptiles are able to attain.
5. The sauropods, because of the relatively small surface-area-to-volume ratio imposed by their immense size, may have had problems shedding excess body heat. This would almost certainly have prevented them from having had a high (avian or mammalian) level of metabolism.
6. The smallest theropods, like *Compsognathus* and *Dromaeosaurus*, were too small to have been inertial homeotherms. They possess skeletal adaptations that indicated high levels of activity and it is possible that some of them may have been endothermic. Possession of feathers in the closely related earliest bird, *Archaeopteryx*, is persuasive evidence for endothermy.

What about the brains of dinosaurs? In what way do brain estimations obtained from fossils help to provide insights into thermoregulation of the dinosaurs? Paleoneurological studies have provided the reconstruction of endocranial casts from brain cases found in fossilized dinosaurs. Unfortunately, endocasts cannot provide a clear picture of the morphology of the brain beyond the masks left in the brain cases. Nonetheless, general features of a dinosaur brain are similar to those of the typical reptilian brain. Moreover, as shown in figure 2.10, the dinosaur brain is consistent with the brain/body size relation of living reptiles.

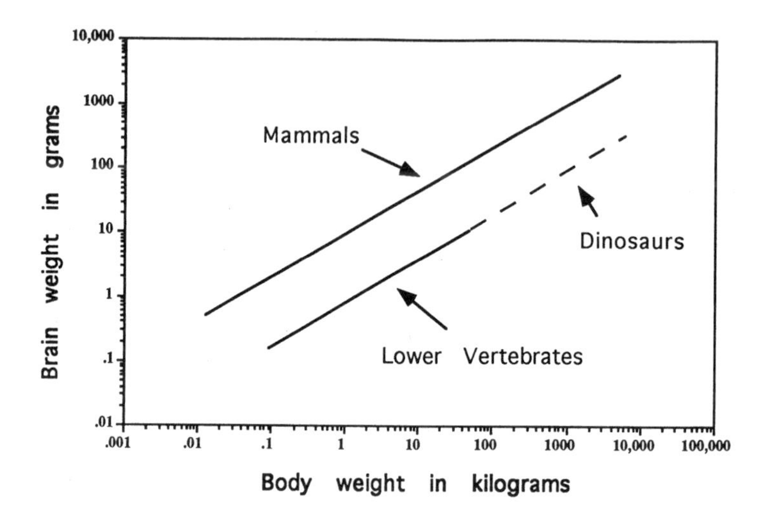

Figure 2.10
Relationship between body mass and brain mass of dinosaurs and lower vertebrates (amphibians and reptiles), compared with mammals. (Modified from Giffin 1994.)

From endocranial casts, only the external morphology can be analyzed and some brain functions be suggested, particularly those processing sensory information. Anatomic and functional estimations of the cerebral cortex, pituitary gland, olfactory bulbs, and optic lobes have been reported (Giffin 1994).

Jerison (1973) reported the endocranial volumes, body volumes, and expected brain volumes of 10 dinosaurs (table 2.2). The expected brain volumes are the result of applying the equation

E (expected brain weight) = k•P (body weight)$^{2/3}$.

The average constant k for fish and reptiles is 0.007. Brain volume (ml) = brain weight (grams). Therefore:

$E = 0.007 \bullet P^{2/3}$.

In reptiles, it has been estimated that only about half of the endocranial space is filled with the brain. Therefore, the actual brain volume of dinosaurs would be half of the endocast volume.

To estimate the relative brain size of an animal compared with other living animals, Jerison (1973) defined the concept *encephalization quo-*

Table 2.2
Quantitative brain and body relations in ten dinosaurs

Genus	Body volume (metric tons)	Endocast volume (ml)	Expected brain volume (ml)
1. Allosaurus	2.3	335	120
2. Anatosaurus	3.4	300	160
3. Brachiosaurus	87.0	309*	1400
4. Camptosaurus	0.4	46	38
5. Diplodocus	11.7	100	360
6. Iguanodon	5.0	250	200
7. Protoceratops	0.2	30	24
8. Stegosaurus	2.0	56*	110
9. Triceratops	9.4	140	310
10. Tyrannosaurus	7.7	404	270

* Reported direct measurement.
Modified from Jerison 1973.

tient (EQ). This is the ratio of actual brain weight to expected brain weight. The expected brain weight is an "average" for living species that takes into account body weight. Consider the following example from Jerison. A squirrel monkey with a body weight of 1000 grams has a brain weight of 24 grams. What is its expected brain weight?

$$E \text{ (expected brain weight)} = k \bullet P \text{ (body weight)}^{2/3}$$

$$E = 0.12 \bullet 1000^{2/3} = 12.$$

In this example, the value of k, as an average for primates, is not 0.007 (used for reptiles) but 0.12. The encephalization quotient (EQ) for the brain of the squirrel monkey is

$$EQ = E \text{ (actual brain weight)} / E \text{ (expected brain weight)}$$

$$EQ = 24/12 = 2.$$

Therefore, the brain of the squirrel monkey is double the brain size of an average living mammal of the same body weight. What about the human brain? The EQ for humans with a brain of E = 1350 grams and a body weight P = 70,000 grams would be

$EQ = 1350/0.12 \cdot 70{,}000^{2/3} = 6.5.$

Thus, the brain of man is six and a half times larger than the average brain for a typical mammal of the same body weight. Humans also have the highest relative brain size of all living vertebrates.

If we return to the brain of the dinosaurs, EQ for *Tyrannosaurus* is

E (actual brain weight) = 530/2 = 265 grams

E (expected brain weight) = $0.007 \cdot 7{,}700{,}000^{2/3} = 273$ grams

$EQ = 265/273 = 0.9.$

In another example, the expected brain weight for the *Diplodocus* would be

$EQ = 50/0.007 \cdot 11{,}700{,}000^{2/3} = 0.1.$

These calculations indicate that the *Diplodocus* brain (herbivore) was about ten times smaller than the average reptilian brain, whereas that of *Tyrannosaurus* (carnivore) was approximately equal to the average brain of a reptile of similar weight. These observations suggest that the herbivorous dinosaurs had smaller brains than carnivorous dinosaurs. The average encaphalization quotient for the ten dinosaurs listed in table 2.2 is 0.56, which is half the expected brain size of reptiles with a mean body weight of 13 metric tons.

But what does this say about the thermoregulatory behavior of these creatures? Certainly very little by itself, but together with other evidence provided by studies on current living reptiles, it supports the idea that dinosaurs, at least the big ones, were ectothermic, most probably ectothermic homeotherms.

In summary, from the vast number of observations that have been made on dinosaurs, it seems that their pattern of behavior was similar in significant ways to those of living animals at a comparable level of brain and skeletal evolution. This assumption is supported by research performed on large living reptiles. In fact, studies on alligators and large turtles have shown that these animals are in fact ectothermic, although they can have warm blood because of their enormous size (Spotila 1980).

Furred Creatures in the Darkness: Thermal Thoughts for Freedom

Romer (1970) pointed out that the reptilian ancestors of mammals were of the subclass Synapsida. These animals diverged very early, at the base of the family tree of that class (figure 2.11). Among the synapsids, the therapsids were the early mammal-like animals in the late Permian and Triassic periods (figure 2.12).

The characteristics of these animals—active, four-footed runners with elbows and knees swinging in toward the body, together with circulatory and respiratory improvements—gave them greater speed. At this time, the diaphragm appeared. Other changes, such as thyroid-regulated and catecholamine thermogenesis, could have occurred during this period.

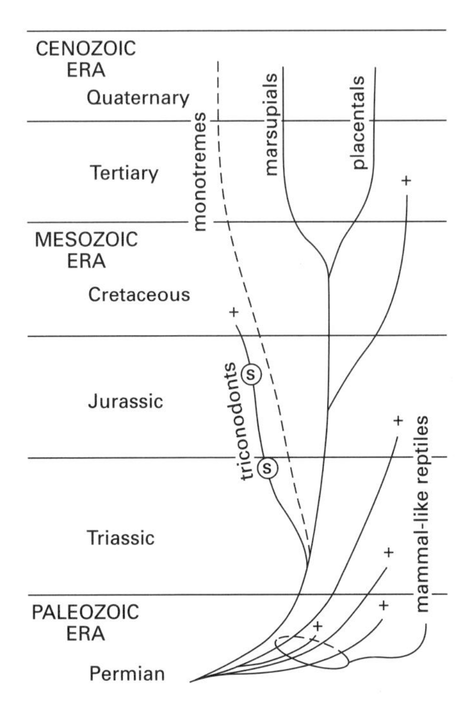

Figure 2.11
Mammalian phylogeny. (Taken from Romer 1970.)

Figure 2.12
A mammal-like reptile. (Taken from Romer 1970.)

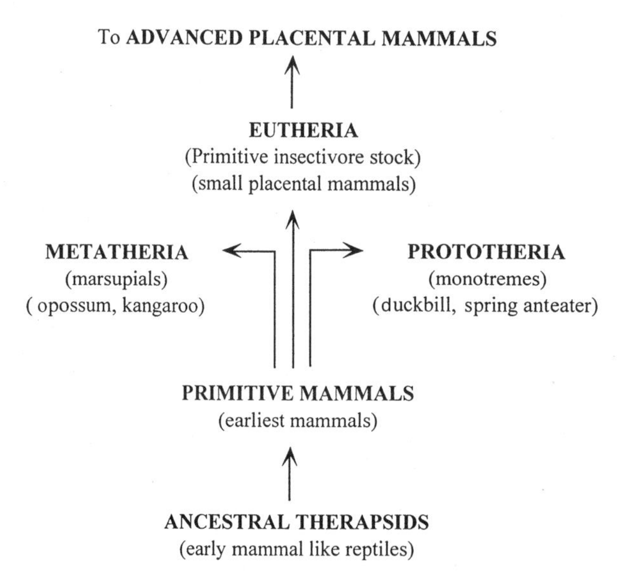

Figure 2.13
Schematic diagram showing family tree of mammals from ancestral therapsids. (Modified from Romer 1970.)

These anatomical and physiological improvements supposedly made the therapsids better suited for higher oxygen uptake than their predecessors. They had a sprawled-limb pose to support their presumed progressive increase in oxidative-type skeletal muscle. Moreover, their postural muscles gave them greater muscle tension at rest than amphibians and reptiles. As stated by Heath (1968 p. 268), "As a result of a shift in posture, then, mammal-like reptiles probably acquired accidentally a system to provide heat even at rest." These characteristics were prerequisites for animals to develop endothermy.

We know today that muscle tension in mammals provides about 30 percent of the heat production under basal conditions (Jansky 1962) and that, compared with extinct reptiles, living mammals have increased pulmonary and cardiovascular function and blood oxygen-carrying capacity, elevated tissue vascularization, and high levels of myoglobin and oxygen extraction rates (Withers 1992). Since very early reptilian evolution, and for millions of years, the mammal-like reptiles (therapsids) from which true mammals evolved (figure 2.13) lived with their contemporaries, the dinosaurs. By the late Triassic period (the era of the dinosaurs), mammal-like reptiles were not able to compete with large reptiles, and therefore their survival became problematic.

The small therapsids, the true predecessors of mammals, survived because they invaded new niches away from those of the reptiles. Jerison (1973) suggested that these new adaptive zones were nocturnal and those conditions would have exerted a strong selection pressure for developing mechanisms to control temperature in the absence of sun light.

These early mammals were an order of magnitude smaller than their precursors, the advanced mammal-like reptiles (30–40 g body weight) (Crompton et al. 1978). The largest ones were probably no more than 60 cm long and weighed about 5 kg. In fact, Romer (1970) suggests they were similar to rats and mice, occupying niches like those occupied by these rodents today (figure 2.14).

Paleontologists have speculated that primitive mammals and their mammal-like reptilian ancestors maintained a constant body temperature. In fact, endothermy was always considered an essential physiological function for animals to survive in nocturnal niches where heat from the sun was absent.

Figure 2.14
Reconstruction of an early mammal (*Megazostrodon rudmerae*) from the late Triassic period, about 200 million years ago. (Taken from Crompton et al. 1978.)

Crompton and colleagues (1978), based on fossil records and comparison of energetics of insectivores, monotremes, and marsupials, have proposed that homeothermy in primitive mammals was acquired in two steps. First came the acquisition of a constant body temperature that was 10°C lower than 37°C. Two families of insectivores, the Tenrecidae and Erinaceidae, that live in nocturnal niches and appear to have lived there since primitive times, have a mean resting body temperature of about 29.5°C (range of 28.5°–30.5°C) at ambient temperatures between 21° and 24°C (Crompton et al. 1978). Also, the energetics (basal metabolic rates) of these insectivores seem to be reptile-like, being less than 1/3 of that predicted for a mammal (Crompton et al. 1978). In contrast, monotremes and marsupials, who had diurnal ancestors, had a basal metabolic rate similar to those of present-day mammals. Crompton et al. (1978 p. 334) suggest that "a low body temperature coupled with a low metabolic rate restricted the first mammals to their nocturnal niches enabling them to survive on much less energy." The second step would have been the acquisition of a higher metabolic rate by some of these nocturnal mammals.

The hypothesis proposed by Crompton and colleagues (1978), based on physiological research performed on extant mammals, seems plausible. However, physiology provides no direct evidence for the real origin of endothermy, which occurred almost 200 million years ago. In a review on the evolution of endothermy in mammals and birds, Ruben (1995 pp. 82–83) stated:

> Virtually all previous interpretations of the metabolical status of extinct taxa have centered on speculative and/or circumstantial evidence, including predator-prey ratios, fossilized trackways, fossil bone oxygen isotope ratios and paleoclimatological inferences, or on correlations with mammalian or avian morphology, such as posture, relative brain size, and bone histology. These arguments are equivocal at best. Furthermore, the majority of the morphological arguments used previously, including specially bone histology data, are based predominantly on apparent similarities to the mammalian or avian condition, without a clear functional correlation to distinctly endothermic process . . . . this situation has changed with the discovery that the nasal respiratory turbinate bones in mammals and possibly birds, are tightly and causally linked to high ventilation rates and endothermy in these taxa.

That mammals and birds, but not reptiles, have special anatomical arrangements in their nasal cavities (turbinates) can be traced in fossil records. This is a fascinating story. From these records, physiological inferences for endothermy have been made. The turbinates are highly complex structures located in the nasal cavities that provide a countercurrent exchange mechanism to preserve heat and water. Thus, during inspiration, cool external air enters the respiratory tract and takes up heat and moisture from the turbinate linings. During expiration, this process is reversed and warm air is cooled as it passes over the turbinates. In this way water and heat are preserved in the organism. These turbinates are present in the nasal cavities of almost all mammals and birds, but are completely absent in all extinct reptiles (including dinosaurs) (figure 2.15). Therefore, it seems that turbinates evolved in parallel with the evolution of mammalian and bird endothermy. Of enormous interest are the studies on cranial fossils which showed that the bony structures to which turbinates attach were present in some therapsids and early mammals. This suggests that some advanced mammal-like reptiles already had high ventilation rates, high metabolic rates, and, supposedly, endothermy—a hypothesis advanced by Van Valen (1960) but not shared by Heath

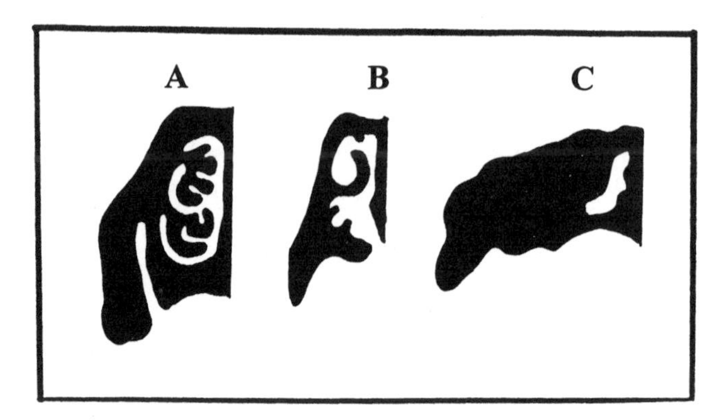

Figure 2.15
Turbinates in mammals (A), birds (B), and reptiles (C).

(1967), and inconsistent with the findings obtained from the brains of therapsids.

But what has the brain to do with this? When trying to explain the achievement of endothermy by primitive mammals, only body changes such as muscle activity, circulatory improvement, metabolic activity, and changes in the surface of the body to diminish heat loss (including the development of hair or fur in the skin) have been considered. However, for mammals to have achieved a constant body temperature, a balance between heat production and heat loss had to be maintained and controlled by the brain, a capacity that has not been achieved in ectothermic animals (fish, amphibians, and reptiles). What do we know about the brain of therapsids and primitive mammals compared with the brain of their ancestors, the reptiles? Is there a significant anatomical difference in brain morphology that could justify the appearance of endothermy? If so, what in the brain changed to allow small mammals to become endothermic?

Heath (1968) believes that the change in limb position in therapsids led not only to the development of endothermy but also to fundamental changes toward reorganization of the brain, in both the motor and the somatosensory areas. In therapsids, changes in the cerebellum have clearly been observed that fit well with better motor coordination and finer motor control in these animals. The fact that the somatotopy of the pyramidal system in the cerebral cortex is found in marsupials, monotremes, and placentals also suggests a common ancestor, the therapsids (Lende 1964). Moreover, we now know that a classical neurochemical pathway—the nigrostriatal pathway, which is involved in motor behavior—plays a role in thermoregulation. Was this pathway involved in the maintenance of motor control, as well as in the control of behavioral thermoregulation, in these reptiles? It is not too speculative to suppose that the most fit animals, transferred these capacities from generation to generation. Moreover, a change in the somatosensory capacities of therapsids fits well with the shift in posture that occurred in these animals, because they were in a position to access more information from the environment.

Despite all of these considerations, however, available evidence from studies of endocasts in therapsids shows that they do not have mammalian-like features in their brains, and that their brains look reptilian in both size and shape. Only from the most primitive mammals is there evidence that they had brains four or five times bigger than those of reptiles of comparable body size. In fact, Mesozoic mammals already had larger brains compared with their very early primitive ancestors, the reptiles and the most immediate mammal-like reptiles, the therapsids. The major features of the endocasts from a therapsid, the mammal-like reptile *Nythrosaurus*, shows that the cerebral hemispheres are long and narrow. In a more primitive mammal, the *triconodon*, the cerebral hemispheres are expanded laterally and dorsally (Jerison 1973). Jerison (1973) shows that the fossil data from triconodon, and others, fall into the polygon for living mammals. Taken together, these data suggest that the brain of therapsids remained reptilian and thus support the view of Heath (1968) that the central integration of thermal information in therapsids remains reptilian.

These observations suggest that endothermy appeared at a time when the brain reached a certain size. But what does endothermy have to do with the enlargement of the brain? Was the achievement of endothermy a powerful determinant for the enlargement of the brain experienced by the early mammals? Or was it just the contrary? This is an unsolved problem of primary importance that has not been appreciated.

The only thing that can be said is that the brain of early mammals was larger than the brain of their predecessors, the reptiles. From the analysis of brain cases, very little can be discerned about the internal configuration of the brain in terms of areas, nuclei, and their connections that could reveal the essential changes in the brain of a mammal compared with a reptile for controlling brain temperature.

One approach to elucidating this problem would be to compare the internal histological structures of the hypothalamus and limbic system of living reptiles with those of living mammals. Any significant difference could be relevant for understanding the neural basis of temperature control in living animals and retrospectively shed some light on the evolution of these parts of the brain in relation to the appearance of endothermy. In any case, it is almost certain that the adaptations of the small reptile-like mammals to life at night would have placed extraordinary demands not only on their temperature-regulating capacities but also on their sensory systems (Jerison 1973). This would have important implications for the evolution of their brains. Speculations about sensory adaptations in nocturnal niches, such as hearing and smell, with regard to enlargement of the brain in early mammals have been offered by Jerison (1973).

After the extinction of big reptiles at the end of the Cretaceous period, mammals equipped with new, powerful capacities for processing sensory information; more efficient locomotor, cardiovascular, and respiratory systems; a high metabolic rate; and fur to minimize the absorption of solar radiation and nightly loss of body heat (Hammel 1976) started the long journey to many different diurnal ecological niches.

The Warm Flight to the Air

Thermoregulation in birds is yet another story, although in some aspects it is very similar to that of mammals. Like mammals, birds are homeothermic endotherms and evolved from reptiles. However, birds evolved from a different stock of reptiles around 30 million years after the appearance of the most primitive mammals. Jerison (1973) suggested that the niche of the earliest birds was often deprived of much of the sun's warmth. Like the appearance of fur in mammals, the evolution of feathers (insulation) seems to have preceded the evolution toward endothermy (Cowles 1946). Furthermore, as occurred in mammals (upright posture and exercise) "incidental" behavior in birds (flight) and excess production of heat could have been at the origin of endothermy. If these considerations are correct, similar selection pressures (deprivation of heat from the sun) and mechanisms for heat production (exercise) for mammals and birds led to the development of a similar thermal strategy: endothermy.

As with mammals, the appearance of endothermy in birds was correlated with the appearance of an increase in brain size compared with that of reptiles having the same body size (Jerison 1973). The brain of primitive and living birds is enlarged relative to that of reptiles. Therefore, it looks as if increased brain size and endothermy may have evolved in parallel fashion among birds and mammals.

An intriguing feature and major difference between thermoregulation in birds and mammals is the central nervous system area responsible for driving autonomic and behavioral responses. In mammals, the hypothalamus and its related limbic structures are the major areas integrating behavioral and autonomic responses for thermoregulation. In birds, autonomic responses seem to be driven by the spinal cord, whereas behavioral responses are driven by the hypothalamus (Rautenberg et al. 1972; Schmidt 1978). For example, heating the hypothalamus of the pigeon fails to evoke a strong autonomic response, but elicits a marked behavioral response. In birds with a naked head and neck, brain temperature can fall 1°C in the cold but rise 2°C during exposure to intense radiant heat (Crawshaw et al. 1990).

3

The Mechanics of Our Environmental Independence: Building Circuits and Chemicals in the Brain

Integrating Neural Mechanisms for Thermoregulation

Although fish, frogs, and reptiles do not have a high metabolic rate and require solar radiation to raise body temperature, they do thermoregulate. Like mammals, these lower vertebrates possess a spinal cord, brain stem, and hypothalamus. When these areas of the brain are stimulated or destroyed, disturbances in the capacity of these animals to thermoregulate occurs, thus indicating that these structures do participate in the control of body temperature. However, these animals are unable to maintain the fine control of core temperature observed in mammals.

In the preceding chapter, we noted that the brain of mammals is significantly larger in relation to body weight, compared with brain size of their predecessors, the lower vertebrates. Is the achievement of homeothermy in mammals linked in some way to brain size? From lesion studies, we know that the mammalian hypothalamus is the primary area within the central nervous system (CNS) controlling body temperature. Is there a significant change in the structure of the hypothalamus of mammals, compared with lower vertebrates, to explain the narrow range within which mammalian body temperature is maintained? Is the hypothalamus in mammals the sole structure controlling body temperature? What do we know of the chemical substances, neurotransmitters and neuromodulators, within thermoregulatory pathways of the mammalian brain that control body temperature?

Until the mid-1960s, the hypothalamus was considered by most investigators in the field to be the only structure able to detect core body temperature. This concept was abandoned when Simon and his colleagues (see Simon et al. 1986 for review) demonstrated that the spinal cord is capable of eliciting all of the thermoregulatory responses evoked by the hypothalamus. Subsequent studies provided evidence for other areas of control, and Satinoff (1978) suggested that the control of body temperature by the CNS in higher mammals was acquired, during evolution, through the development of different hierarchically arranged parallel systems. Rather than a single integrator with multiple inputs and outputs, the proposed model includes integrators for all thermoregulatory responses at several levels of neural control that can facilitate or inhibit upper and lower levels of control. In fact, considerable evidence points to the involvement of many areas of the CNS in the control of body temperature (Hardy 1961; Zeisberger 1987). If this theory is correct, it would imply that a sudden jump from poikilothermic to endothermic animals did not occur, in the sense that endotherms acquired a unique set-point absent in poikilotherms. It suggests, instead, that throughout evolution, all living systems had inputs, thermostats, and outputs, however simple, and that nature developed more and more sophisticated systems superimposed on those already developed. These ideas follow the Jacksonian concept of evolution that nature superimposes new neural structures on preexisting systems rather than replacing them. The added structure would control the existing structure. In mammals, the hypothalamus usually controls the activity of thermoregulatory effector mechanisms along the neuraxis. This latter arrangement, reached in primitive mammals and birds, was probably so efficient that it was retained without modification for about 150 million years.

An example of this process would be the following senario. A high internal heat production was proposed by Heath (1968) as a by-product of primitive reptiles changing posture from a sprawling position to standing erect. However, as the capacity to produce internal heat improved, it probably became a thermal stress over time, and required the development of *internal* thermodetectors and a mechanism of heat loss. Because primitive mammalian ancestors already possessed a very effective vasomotor system and breathing apparatus, thermosensors only needed to gain control over these existing systems to provide adequate heat transfer and heat loss via panting.

Experimental evidence has shown that mammals have many thermoregulatory systems, including the spinal cord, midbrain, hypothalamus, and other structures in the limbic system (Chambers et al. 1974; Liu 1979)(see also chapter 4). At all levels, a set-point range exists that is widest in the spinal cord and becomes more narrow in the hypothalamus. Thus, the hypothalamus exerts the highest level of control. Figure 3.1 shows the width of the range of regulation in spinal, decerebrate, hypothalamic, and normal intact mammals. Bligh (1966) proposed that homeotherms have a "fine-tuned" temperature control as well as "broadband" control. Both are located in the hypothalamus, but ectotherms have only broadband control. He further suggests that the fine-tuned control may have evolved independently in different mammalian orders without a common origin.

Ontogeny and Phylogeny of Thermoregulation

If the concept of multiple integrated thermoregulatory systems is correct and if Haeckel's hypothesis that ontogeny replicates phylogeny is correct, then the idea of a sharp, clear-cut distinction between ectothermic and endothermic animals should be abandoned. Thus, what evolved were animals that adapted different thermoregulatory strategies along with more and more sophisticated neural control systems. This was accomplished by different bodily transformations. This transition was characterized in phylogeny by poor regulation (wide range of body temperature; fish, amphibians, reptiles) and more fine regulation (narrower range of body temperatures; birds and mammals). Examples of how ontogeny simulates phylogeny in thermoregulation is found in the maturation of the CNS in infants and during the acquisition of adult temperature regulation (see chapter 4). Also, functional disruption of the hierarchical integration among temperature controllers along the neuraxis is produced in adult mammals during sleep (see chapter 4). Figure 3.2 compares the range over which temperature is regulated after placing lesions at different levels of the neuraxis in a mammal (A) with the range of temperature at which different species thermoregulate (B).

As mentioned in the preceding paragraphs, all vertebrates, from fish to mammals, have thermoregulatory systems located in the spinal cord,

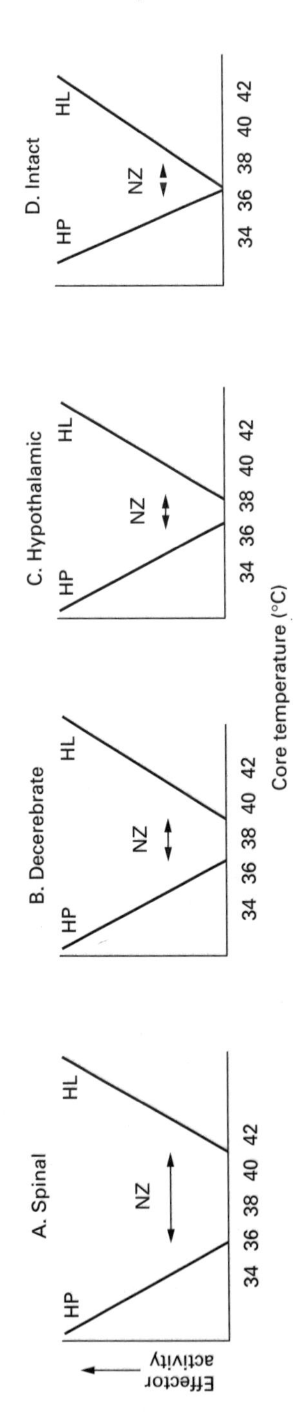

Figure 3.1
Thresholds for activating heat production and heat loss at different hierarchical sites of control. (Modified from Satinoff 1978.)

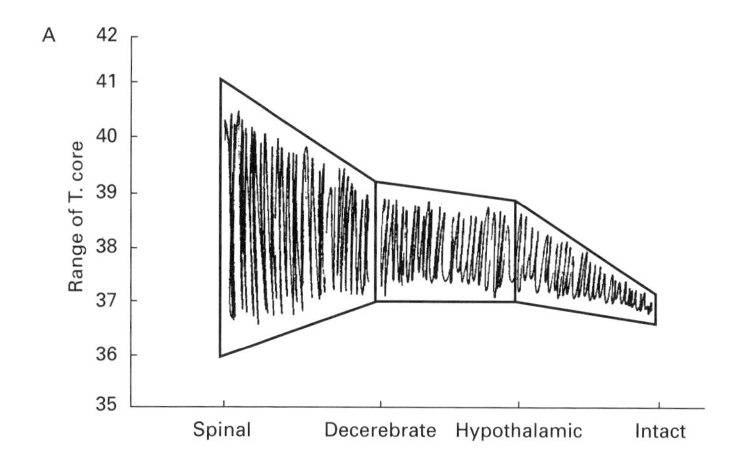

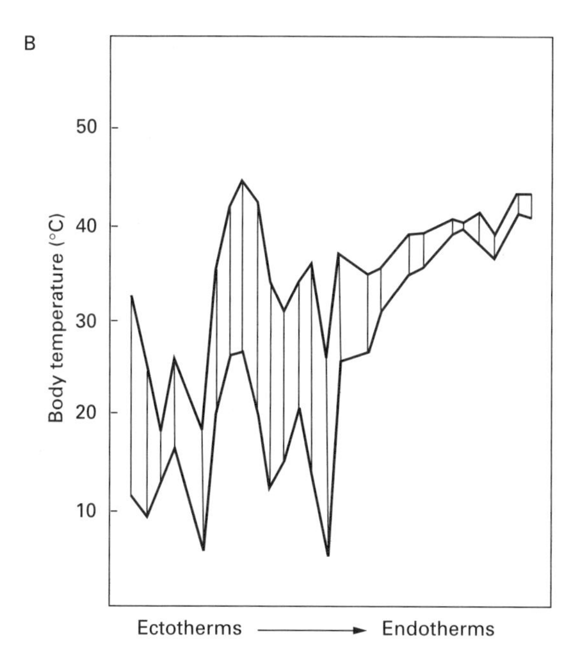

Figure 3.2

(A) represents the range of variation at which a spinal, decerebrate, hypothalamic, and intact animal thermoregulate. A trend from a wide to a more narrow range of temperature regulation is illustrated when ascending the neuraxis from the spinal cord of the intact animal. The narrower range of temperature regulation in the intact animal probably involves a finer and further control by other structures in the limbic system. This general trend could be taken as an indication of the ontogenic development of brain control mechanisms of temperature regu-

brain stem, and hypothalamus, but only birds and mammals are homeotherms. Since the mammalian hypothalamus exhibits large anatomical differences from the hypothalamus of all other vertebrates, and since the hypothalamus is the seat of higher and finer temperature control, would comparative cytoarchitectural studies of the hypothalamus provide insight into the development of homeothermy?

The Hypothalamus of the Mammal

Before analyzing the gross differential morphological characteristics of the hypothalamus through evolution, let's review the structure of the mammalian hypothalamus. Many studies have described the nuclei and boundaries of the hypothalamus in the mammal, both in the adult and in various embryological stages (Haymaker et al. 1969). A detailed description and discussion of the morphological and embryological criteria on the basis of which the different regions and subdivisions of the mammalian hypothalamus have been established can be found in Christ (1969).

For the purpose of this discussion, the hypothalamus of the mammal can be divided into the preoptic-anterior hypothalamic area, the tubero-infundibular or middle hypothalamus, and the more caudal part or posterior hypothalamus. All three regions can be further subdivided into medial, lateral, dorsal, and ventral parts. In humans, the following main nuclei can be identified (Mora and Sanguinetti 1994)(figure 3.3):

Anterior Hypothalamus

1. Paraventricular (dorsal) nucleus and supraoptic (ventral) nucleus
2. Medial preoptic nucleus (dorsal)
3. Anterior hypothalamic nucleus (ventral)
4. Lateral preoptic nucleus (ventral)

regulation in mammals. (Redrawn from figure 3.1.) (B) shows the transition from poikilothermic to endothermic animals and the corresponding narrowing of the range of temperature regulation. This could be interpreted as a gross indication of the phylogenetic development of temperature regulation during evolution. (Redrawn from figure 2.1.)

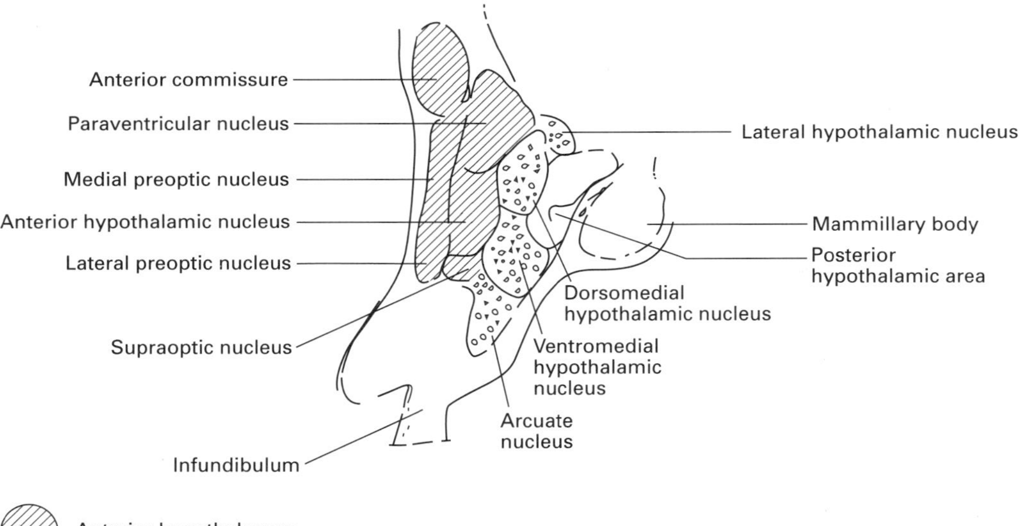

Figure 3.3
Sagittal section of the human hypothalamus showing the main nuclei in the anterior, medial, and posterior hypothalamus. (From Mora and Sanguinetti: 1994.)

Middle Hypothalamus (Tubero-Infundibular)

1. Dorsomedial hypothalamic nucleus (periventricular)
2. Lateral hypothalamic nucleus (dorsal)
3. Ventromedial hypothalamic nucleus (periventricular)
4. Tuber nuclei (ventral)

Posterior Hypothalamus

1. Posterior hypothalamic nucleus (dorsal)
2. Mammillary nuclei: ventral, intercalary, and lateral (ventral)

For comparative gross anatomical purposes, this brief account of the main nuclei of the hypothalamus is sufficient as a reference. In fact, only the three main parts of the hypothalamus will be comparatively analyzed through evolution. Moreover, despite the fact that the hypothalamus is an area of the brain that integrates the function of many brain areas (brain stem, limbic system) and information from peripheral and central thermoreceptors, as well as vegetative input, neuropharmacologists and neurochemists often speak in such vague terms as "anterior hypothalamus" and "posterior hypothalamus."

The Hypothalamus Through Evolution: From Fish to Mammal

It should be said at the outset that no comparative studies exist to allow us, even in the simplest terms, to trace a homology among the different nuclei in the hypothalamus of different classes of animals. The concept of homology is still controversial when referring to the same nuclei in different animals during the process of trying to draw phylogenetic conclusions (Christ 1969). Nonetheless, some general conclusions and speculations have been made (Crosby and Woodburne 1940; Crosby and Showers 1969; Clark 1979) (figure 3.4).

Early in phylogeny, the hypothalamus was an arrangement of central gray surrounded by a periphery of scattered cells and fibers. This central gray matter in cyclostomes (petromyzonts) can be separated into a periventricular area (the nucleus periventricularis preopticus) and its surrounding border. In fish, this border of cells consists of the medial and lateral preoptic and hypothalamic areas. This simple pattern of a periventricular preoptic nucleus and a medial and lateral preoptic area

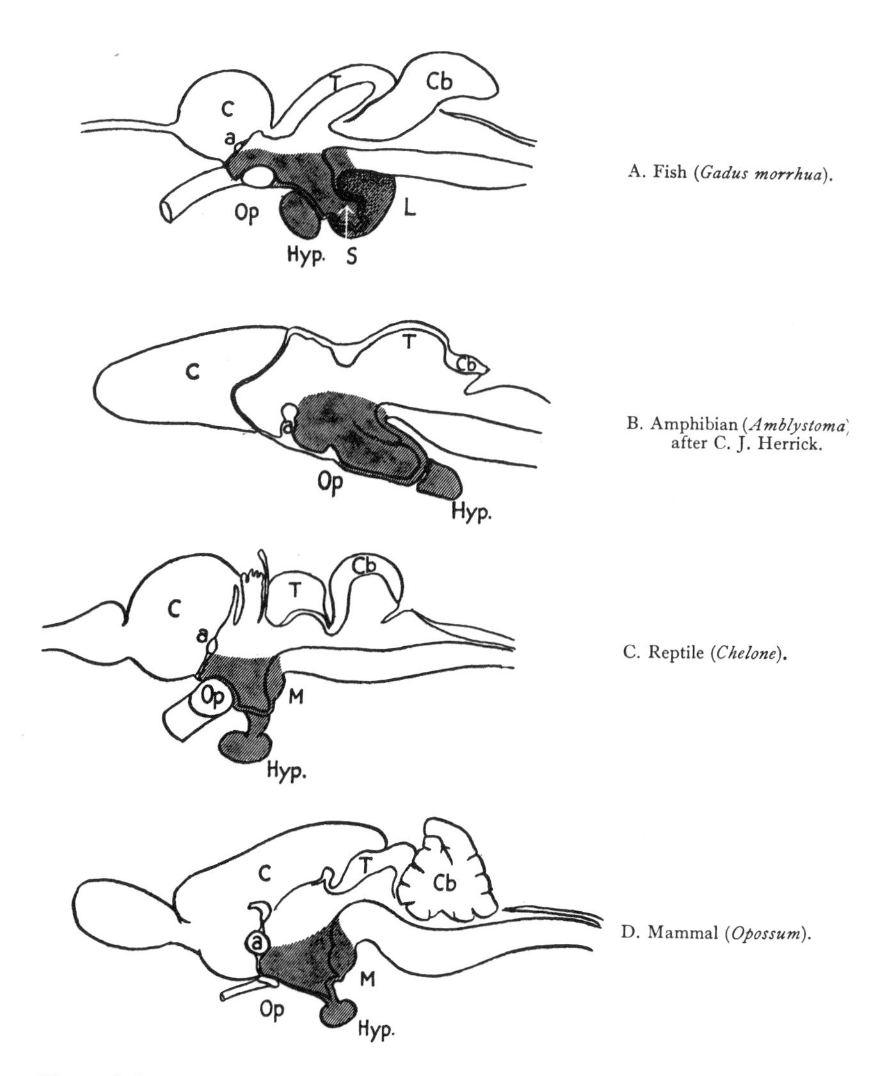

Figure 3.4
A systematic view of the extent of the hypothalamus through the saggital sections of the brains of a fish, an amphibian, a reptile, and a mammal. a, anterior commissure; C, cerebral hemisphere; Cb, cerebellum; Hyp, hypophysis; L, lobus inferior; M, mamillary body; Op, optic chiasm; S, saccus vasculossus; T, tectum of midbrain. (From LeGros Clark 1938.)

can be recognized among all vertebrates through the phylogenetic series. In fact, the lateral preoptic area constitutes the preoptico-hypothalamic transition area (in fish called the nucleus preopticus). Of particular relevance is the constancy of the pars magnocellularis of this nucleus preopticus (magnocellular preoptic nucleus), although it presents great variations with respect to size and degree of differentiation in different fish. In some fish, as in amphibians, the nucleus magnocellularis shows a tendency to split into dorsal and ventral parts, which are seen more clearly in reptiles. These two parts of the nucleus magnocellularis correspond to the paraventricular and supraoptic nuclei in mammals. The medial preoptic area has little representation in fish, and has even less representation in cyclostomes.

In fish, the middle and posterior regions of the hypothalamus in their medial positions have a periventricular and a medial hypothalamic area. In cyclostomes, there is little differentiation of the hypothalamus caudal to the preoptic-anterior hypothalamus, and the medial region, in particular, is very poorly developed at this level (middle and posterior region).

In some fish, a primordial ventromedial hypothalamic nucleus can be vaguely identified, and in the caudal extremity of the hypothalamus, what could be the primordium of the mammillary bodies has been identified. The cells in this region have distinct characteristics and are connected to the thalamus; in higher vertebrates (mammals) this corresponds to the mamillo-thalamic tract. However, this bundle of fibers appears only in mammals and is rudimentary in a few reptiles.

In amphibians, the preoptic-anterior hypothalamus does not present any major differences from the hypothalamus of fish. In fact, "in some respects the hypothalamus of amphibia is not as complicated in its cytoarchitecture as in fish, and its relative simplicity suggests a retrograde process in the evolutionary development of this class of vertebrates" (LeGros Clark 1938). In lower tailed amphibians, a periventricular preoptic nucleus and a magnocellular preoptic nucleus can be distinguished. However, a fully developed medial preoptic area is not present. In tailless amphibians, this structure can be recognized, but it is poorly developed. Figure 3.5 shows a sagittal section of the brain of a tailless amphibian (*Bufo vulgaris*) in which the two parts (parvocellular and magnocellular of the nucleus preopticus) can be recognized. As in fish, only the pri-

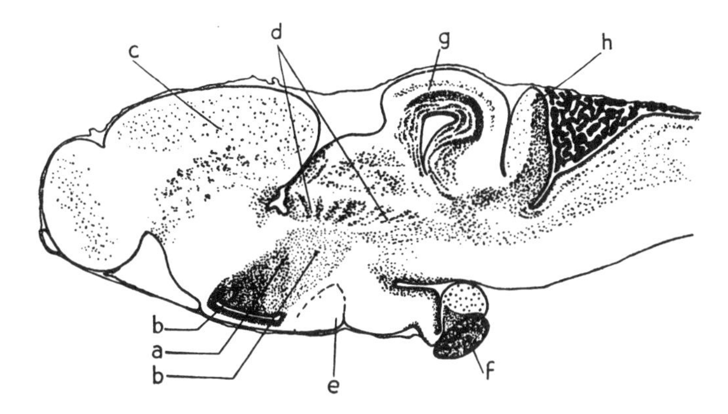

Figure 3.5
Sagittal sections of the brain of a frog (*Bufo vulgaris*). a: pars magnocellularis of the nucleus preopticus. b: pars parvicellularis. c: optic tract. d: hypothalamus. e: hypophysis. (From LeGros Clark 1938.)

mordium of the ventromedial nucleus has been identified. In tailless amphibians, at the caudal end of the hypothalamus, some gray matter has been regarded as mamillary bodies.

In summary, the portion of the hypothalamus that relates to the preoptic area of fish and amphibians in position and function forms the rostral hypothalamic area of mammals. The dorsal and ventral hypothalamic regions caudal to the preoptic area show no separation into tuberal and mammillary areas in tailed amphibians, except as this is foreshadowed by fiber connections. However, in passing from less differentiated submammals such as tailed amphibians to more specialized reptiles and birds, there is a gradual emergence of the centers that characterize the tuberal and mamillary portions of the mammalian hypothalamus.

Reptiles have a more advanced and developed hypothalamus that contains the three main subdivisions of the hypothalamus (as defined in mammals). These are recognized without difficulty, although they have much less cellular differentiation (figure 3.6).

In many species of reptiles (turtles, snakes, lizards, and alligators), the preoptic recess is well developed. The periventricular nucleus preopticus and a medial preoptic area are contained within the medial part of the preoptic region. The nucleus magnocellularis is divided into two parts, the paraventricular and the supraoptic nuclei. The supraoptic nucleus

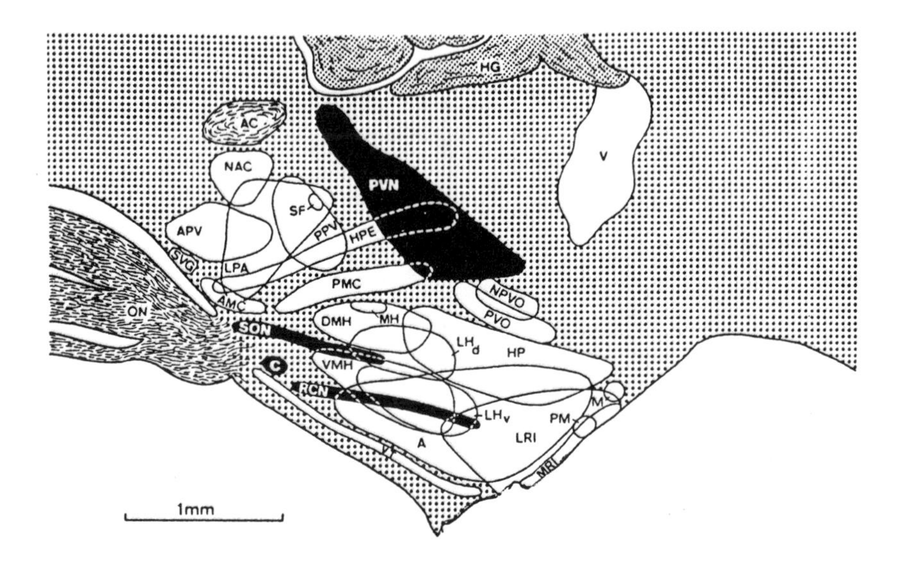

Figure 3.6

Diagrammatic representation of the main hypothalamic nuclei of a reptile (*Naja naja*). A, nucleus arcuatus; AC, anterior commissure; AMC, nucleus microcellularis anterior; APV, nucleus periventricularis anterior; C, nucleus circularis; DMH, nucleus hypothalamicus dorsomedialis; HG, habenular ganglion; HP, nucleus hypothalamicus posterior; HPE, nucleus hypothalamicus periventricularis; LH_d, nucleus hypothalamicus lateralis, dorsal division; LH_v, nucleus hypothalamicus lateralis; ventral division; LRI, nucleus lateralis recessus infundibuli; M, nucleus mamillaris; MH, nucleus hypothalamicus medialis; MRI, nucleus medialis recessus infundibuli; NAC, nucleus commissurae anterioris; NPVO, nucleus of the paraventricular organ; ON, optic nerve; PM, nucleus praemamillaris; PMC, nucleus microcellularis posterior; PPV, nucleus periventricularis posterior; PVN, nucleus paraventricularis; PVO, paraventricular organ; RCN, nucleus retrochiasmaticus; SF, nucleus subfornicalis; SON, nucleus supraopticus; V, third ventricle; VMH, nucleus hypothalamicus ventromedialis; VT, nucleus ventralis tuberis. (Modified from Rao et al. 1981.)

(usually composed of a major periventricular cell group and a pars diffusa) varies greatly in different reptiles in both magnitude and position. Nonetheless, both nuclei can be recognized in most species. In some reptiles, the medial preoptic area is developed to different degrees.

In reptiles, the infundibular region or medial hypothalamus contains two main cells groups, the nucleus ventralis hypothalamicus (homologue to the mammalian nucleus ventromedialis) and the nucleus lateralis hypothalamicus (which, as in mammals, is a bed nucleus through which the medial forebrain bundle sends fibers to and receives fibers from different areas of the brain stem and forebrain).

The most posterior part of the hypothalamus is the reptilian pars mamillaris, located, as in mammals, behind the infundibular region. This corpus mamillare, as well as the mamillo-thalamic tract running dorsally to and within the thalamus, is recognized in all subclasses of reptiles. The posterior hypothalamus is represented by a mass of medium-size cells without distinguishing characteristics that continue without demarcation into the midbrain tegmentum (Crosby and Showers 1969).

The avian hypothalamus, like the hypothalamus of the mammal, presents three well-defined regions: the anterior preoptic hypothalamus, middle,hypothalamus (tuberal), and posterior hypothalamus (mamillary). These three areas, and the nuclei contained within them, are comparable with those in the mammal. Of the three areas, the middle-hypothalamic region is the largest; it is the rostral border marked by the anterior commissure (dorsally) and optic chiasma and supraoptic decussation (ventrally). The posterior border is best estimated by the appearance of the medial mamillary nucleus. The lateral borders are marked by fiber tracts. The posterior hypothalamus (mamillary region) is comparable with the mamillary bodies in mammals, but in birds is characterized by the presence of mamillary nuclei. Kuenzell and van Tienhoven (1982) have concluded that of the three regions, it is most difficult to clearly distinguish hypothalamic nuclei in the posterior hypothalamus. They suggest that fusion of some clusters of cells in the caudal hypothalamus accounts for part of the difficulty.

To summarize the phylogeny of the hypothalamus, the early central hypothalamic gray in primitive vertebrates reveals a rostal part related to the preoptic region and a caudal part divisible into dorsal and ventral regions. In the rostal part, the constancy of the nucleus magnocellularis along phylogeny is one of the most interesting features. The pars infundibularis, middle hypothalamus, develops from the dorsal regions. In fish and amphibians, it is relatively diffuse; only in reptiles do the cell groups resembling those of mammals appear. Also, the mamillary and premamillary nuclei and the posterior hypothalamic area develop from the more caudodorsal region. The mamillary bodies of the hypothalamus are very poorly differentiated in fish and amphibians. In reptiles, it appears as a group of neurons that can be recognized as the pars mamillaris. Only in mammals are the mamillary bodies better defined anatomically.

Is the Posterior Hypothalamus Subserving the Neural Circuitry for the Higher Set-Point Reference in Mammals?

Most intriguing is the late appearance in evolution of the posterior hypothalamic area, which includes the posterior hypothalamic nucleus and the mamillary bodies. Only mammals show the full anatomical characteristics of this hypothalamic area. Has this observation any special significance with regard to the thermoregulatory capacities of the mammal?

Keller (1933) published a series of observations in which he showed that bilateral ablations of the posterior hypothalamus in dogs destroyed the ability of the animals to maintain a stable body temperature when exposed to a cold environment. Moreover, thermal thresholds in these animals were ten times higher than those of normal dogs. The analysis of these initial observations led Hardy (1961) to hypothesize that the function of the posterior hypothalamus was one of integration and transmission rather than of thermal sensitivity. Ranson (1940) described alterations in body temperature in the cat and monkey after discrete lesions of the posterior hypothalamus. Thompson (1959) showed that complete ablation of the posterior hypothalamus of the dog impaired the development of fever. Also, dogs with posterior hypothalamic lesions had febrile responses if exposed to a hot environment, and no response, or even hypothermia, when exposed to a cold environment. These responses were impaired when the entire hypothalamus was destroyed.

Thus, it looks as if complete destruction of the posterior hypothalamus prevents pyrogenic action in most cases. If pyrogens elevate the set-point (which is supported by considerable evidence; see Stitt 1979), these observations would suggest that the posterior hypothalamus is involved in the neural circuitry responsible for the set-point. Additional support for a set-point function within the posterior hypothalamus comes from the hibernation literature. Posterior hypothalamic, but not preoptic anterior hypothalamic (POAH), lesions prevent hibernation (Malan 1966a, 1966b; Satinoff 1967); however, POAH-lesioned animals cannot arouse from hibernation. As a result, they die (Satinoff 1967).

Myers and Veale (1970) reported that the posterior hypothalamus, but not the anterior hypothalamus, was sensitive to changes in extracellular concentrations of Na^+ and Ca^{++} ions for altering body temperature. Gisolfi and Wenger (1984) subsequently concluded that the set-point for thermoregulation was not located in the posterior hypothalamus.

When all of these data are taken together, and when they are considered in light of the late appearance of the posterior hypothalamus and achievement of homeothermy in evolution, perhaps the posterior hypothalamus deserves a more prominent role in temperature regulation than current investigators in the field attribute to it.

Neuronal Model of Hypothalamic Thermoregulation

Thermal information reaches the central controlling areas of the brain via cutaneous thermoreceptors and deep body thermoreceptors. Cutaneous thermoreceptors are stimulated when the skin of the body is warmed or cooled. This afferent information enters the spinal cord, synapses with a second-order neuron, and ascends to the ventrobasal complex and intralaminar nuclei through the anterolateral tract (lateral spinothalamic tract). Collaterals from this pathway reach the reticular formation in the brain stem. From the basal complex, thermal information reaches the cerebral cortex (somatosensory cortex). From the intralaminar nuclei, thermal information reaches the hypothalamus. Other pathways that transmit peripheral thermal information could reach the hypothalamus via the reticular formation in the midbrain. Midbrain neurons responding to warm and cold have been found. Also, neurons in the midbrain can respond to changes in hypothalamic temperature, which suggests that the hypothalamus (preoptic anterior hypothalamus) also sends information back to the brain stem (H. Sato 1984). Serotonin neurons and noradrenergic neurons from the brain stem may participate in conducting or modulating thermal information to the hypothalamus (Hellon 1975; Hinckel and Schroder-Rosenstock 1981).

As proposed at the beginning of this chapter, thermal information is integrated at various levels of the neuraxis in a hierarchically organized fashion: spinal cord, brain stem, and hypothalamus. Presumably an integrator receiving input and transmitting output signals exists at each of these levels. However, it is generally accepted that the hypothalamus is the highest station acquired through evolution for thermoregulation. Moreover, it is in this area of the brain that research has been concentrated and considerable knowledge has accumulated. How are thermal inputs organized in the hypothalamus to activate thermoregulatory responses to heat and cold stimuli? The answer is complex and not entirely known. The first neuronal model of hypothalamic thermoregulation was presented by Hammel (1965). Since then, numerous models have been proposed to explain how a balance between heat production and heat loss is achieved within neural structures of the brain (see chapter 7 for discussion) (Bligh 1972; Boulant 1980; Myers 1980). Figure 3.7 illustrates our current understanding of hypothalamic thermosensitivity and neuronal firing rates in response to peripheral and central thermal stimuli. Important characteristics that should be represented in any neuronal model of thermoregulation are the following: (a) synaptic inhibition by temperature-insensitive neurons of warm-sensitive neurons; (b) warm- and cold-receptors in the POAH receive afferent signals from cutaneous and spinal thermal sensors, which means that a thermosensor can act as an integrator; and (c) changes in peripheral temperature can alter the sensitivity and firing rates of POAH thermosensitive neurons. For example, warming the skin increases heat loss but reduces the sensitivity of hypothalamic warm receptors. On the other hand, cooling the skin decreases the firing rate of warm-sensitive neurons but enhances their sensitivity.

In figure 3.7, Boulant (1996) depicts a model to explain how different groups of temperature-sensitive neurons (based on differences in thermosensitivity) can elicit different thermoregulatory effector responses. He proposes 4 types of thermosensitive neurons; 3 are warm-sensitive (a, b, and c) and 1 is cold-sensitive. Thermosensitive warm-neuron "a" has a

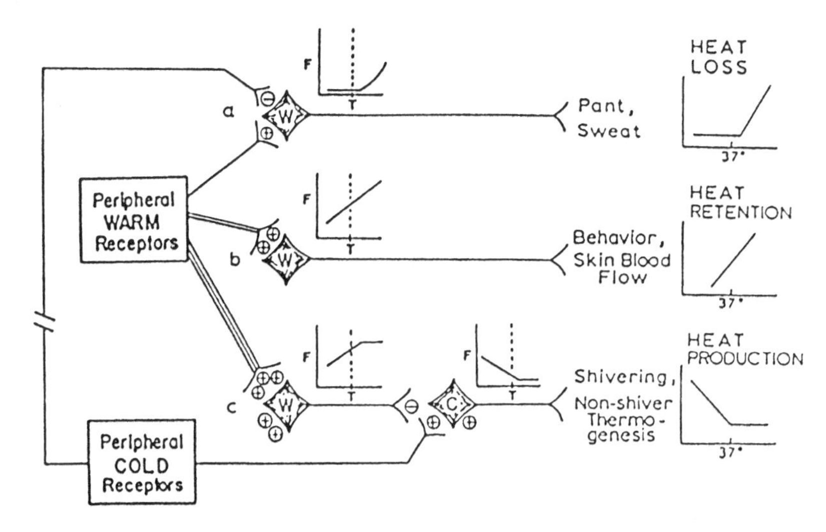

Figure 3.7
A neuronal model proposed by Boulant (1996) in which 4 types of thermosensitive neurons are depicted: warm (W) neurons (a, b, and c) and cold (C) neurons. The cold and warm inputs to these neurons are depicted as excitatory (+) and inhibitory (-). The frequency (F) of neuronal firing is plotted as a function of brain temperature (T).

low spontaneous firing rate and responds only when the animal is hyperthermic (when hypothalamic temperature is above the thermoneutral zone). It is postulated that these neurons most likely control heat loss (sweating or panting). The low firing rate is attributed to inhibition from peripheral cold-receptors and/or limited input from peripheral warm-receptors.

Warm-receptors labeled "b" are thermosensitive above and below normal hypothalamic temperature and therefore are most likely to control heat retention, that is, skin blood flow and behavioral thermoregulatory responses. These neurons receive moderate inputs not only from ascending pathways but also from limbic structures such as the hippocampus, which participates in functions including emotion and motivation, learning and memory (see chapter 4). This is most interesting because it has been suggested that "b" neurons play a role in behavioral thermoregulation (rewards and pains) and skin blood flow (a response that is often influenced by emotion).

Warm-sensitive "c" neurons have the highest spontaneous firing rate. This is attributed to the fact that they receive the greatest amount of afferent excitatory input and may have a greater sensitivity to endogenous factors including testosterone and estradiol, increased osmolality, decreased glucose, and so on. Because these neurons have such a high spontaneous firing rate, it is unlikely that they would increase their firing rate during hypothalamic warming. Thus, Boulant (1996) suggests that these neurons inhibit cold-sensitive neurons, and therefore inhibit shivering and nonshiver thermogenesis.

An important aspect of figure 3.7 is cold-reception and activation of the shivering response. In Boulant's model, a major portion of peripheral input impinges on neurons responsible for heat production rather than heat loss. Cold-sensitive neurons receive direct input from cutaneous cold receptors and indirect input from cutaneous warm receptors mediated through POAH warm-sensitive neurons. The latter neurons exert a negative influence on POAH cold-sensitive neurons. Thus, with a cold skin, POAH cold-sensitive neurons are markedly activated because of positive input from cutaneous cold-receptors and the removal of negative input from cutaneous warm-receptors. This analysis fits well with in vivo studies on experimental animals and studies in humans showing that changes in skin temperature are 2 to 3 times more effective in activating heat production than heat loss (Benzinger et al. 1963; Hellstrom and Hammel 1967).

Transporters of Information: Neurotransmitters and Neuromodulators

Since the mid-1960s, a multitude of studies have reported the effects of numerous putative neurotransmitters, injected peripherally or directly into the brain, on changes in body temperature (Clark 1979). What is currently known about neurotransmitters and neuromodulators in thermoregulation, and what do we know about the molecular events taking place at specific synapses during the process of transmitting thermal information among neurons in the brain? Before discussing the neurochemistry of thermoregulation, we present a brief introduction and synthesis of current knowledge about neurotransmitters and neuromodulators.

Since Otto Loewi (1921) first proposed acetylcholine (Ach) as a neurotransmitter in the vagal terminals of the heart, many neurotransmitters have been proposed to play a role in the CNS. Evidence supporting noradrenaline, dopamine, serotonin, GABA and glutamate as neurotransmitters soon followed that for Ach. These neurotransmitters, once released, act on presynaptic as well as postsynaptic receptors. Pharmacological studies have identified many receptor subtypes for Ach (muscarinics M1, M2, M3, M4, M5, and nicotinics), noradrenaline ($\alpha 1$, $\alpha 2$, $\beta 1$, $\beta 2$), dopamine (D1, D2, D3, D4, D5), serotonin (5HT1, 5HT2, 5HT3, 5HT4, 5HT6, 5HT7), GABA (A and B), and glutamate (AMPA, kainate, NMDA, and metabotropic).

Much has been learned about the molecular machinery responsible for the rapid release of these substances from presynaptic terminals in response to depolarization produced by action potentials. The same can be said about the role of specific transporters located in presynaptic terminals and glia that participate not only in terminating the synaptic action of a neurotransmitter but also in releasing neurotransmitters into the extracellular space (Cooper et al. 1991).

Together with neurotransmitters, more than 50 different neuropeptides have been proposed as candidate neuromodulators in the CNS. Some of the more prominent candidates are listed in table 3.1. Although we do not know the true physiological role of neuropeptides in the CNS, all receptors for neuropeptides currently known are coupled to G-proteins. This supports the idea that neuropeptides play a role as neuromodulators, that is, they do not activate or inhibit neuronal excitability, but modulate the action of neurotransmitters.

One of the most interesting findings since the late 1970s has been the coexistence of different neurotransmitters and neuromodulators in the same neuron (Lundberg and Hokfelt 1985). Most often, this coexistence involves monoaminergic neurotransmitters and different types of peptides. Table 3.2 shows examples of coexistence of neurotransmitters and neuromodulators in neurons in both the peripheral nervous system and the CNS. This coexistence of different transmitter substances in the same neural terminal and their simultaneous release have implications for a real understanding of the synaptic transmission of information (Lundberg and Hokfelt 1985).

Table 3.1
Neurotransmitters and neuromodulators

Neurotransmitters and neuromodulators
*Acetylcholine (ACh)
*∂-Aminobutyric Acid (GABA)
Adenosine and Adenosine Polyphosphates
*Adrenaline (N)
Carbon Monoxide (CO)
*Dopamine (DA)
*Glutamate (GLU)
*Glycine (GLY)
*Histamine
Neuroactive Peptides
*Angiotensin
*Bombesin
*Bradykinin
*Calcitonin Gene-related Peptide (CGRP)
*Cholecystokinin
Galanin
Gastrin
*Glucagon
Insulin
Neurokinin (NKA)
*Neuropeptide Y (NPY)
Neurophysins
*Neurotensin (NT)
Nitric Oxide (NO)
*Noradrenaline (NA)
*Opioid Peptides
*Dynorphin
*ß-endorphin
*Enkephalins
Leucine
Methionine
*Oxytocin
*Secretin
*Serotonin (5HT)
*Oxytocin
*Secretin
*Serotonin (5HT)
*Somatostatin
*Substance P
*Vasoactive Intestinal Polypeptide (VIP)
*Vasopressin

*Produces changes in body temperature when injected into the CNS.

Table 3.2
Coexistence of neurotransmitters and neuromodulators

Transmitter	Peptide	Location
GABA	Enkephalins	Striatum
	Substance P	Striatum
	Somatostatin	Cortical and hippocampal neurons
	Cholecystokinin	Cortical neurons
	Motilin	Cerebellum
Acetylcholine	VIP	Parasympathetic and cortical neurons
	Substance P	Pontine neurons
Norepinephrine	Somatostatin	Sympathetic neurons
	Enkephalin	Sympathetic neurons
	NPY	Medullary and pontine neurons
	Neurotensin	Locus coeruleus
Dopamine	Cholecystokinin	Ventrotegmental neurons
	Neurotensin	Ventrotegmental neurons
Epinephrine	NPY	Reticular neurons
	Neurotensin	Reticular neurons
Serotonin	Substance P	Medullary raphe
	Thyrotropin-releasing hormone	Medullary raphe
	Enkephalin	Medullary raphe
Vasopressin	Cholecystokinin	Magnocellular hypothalamic neurons
	Dynorphin	Magnocellular hypothalamic neurons
Oxytocin	Enkephalin	Magnocellular hypothalamic neurons

Modified from Cooper, Bloom, and Roth 1991.

In fact, since the first neurotransmitters were located in neurons in specific areas of the brain (Dahlstrom and Fuxe 1964), there has been a tendency to correlate a specific neurotransmitter with a specific neural pathway, and this pathway with a single function and a single pathology; for instance, dopamine, the nigrostriatal pathway, motor behavior, and Parkinson's disease. Today, we know that such an interpretation of CNS function is oversimplified. Specific functions in the brain are coded by multiple neurotransmitter systems, neuromodulators, nerve growth factors, hormones, and other chemical messengers in synaptic as well as volumetric dynamic interactions in specific circuits formed by multiple sets of neurons (Brezina and Weiss 1997; Segovia et al. 1997). Pathology arises from a disruption in this physiological interplay produced by numerous agents and circumstances. These neurotransmitter interactions, shown to occur in many different areas of the brain, also occur in the hypothalamus (Exposito et al. 1995).

To complicate matters more, other compounds, among them nitric oxide (NO) and carbon monoxide, can influence the modulation of neurotransmitters' release. These findings have important consequences for our understanding of synaptic function. For instance, the release of NO from a specific set of synapses (100 mu) in the brain could influence around 2 million synapses (Garthwaite 1995).

In summary, an ultimate goal in neurobiology will be to integrate all available information to understand and to predict the function of a specific circuit in the brain. However, the analysis of all the elements involved in the neurotransmitter complexity of a specific circuit is making it more difficult to unravel the real molecular functioning of a system. Although attempts have been made to understand certain circuits in the brain, we are far from understanding the circuitry that controls temperature.

Neurotransmitters, Neuromodulators, and Thermoregulation

Feldberg and Myers (1964) proposed that a balanced release of noradrenaline and serotonin in the hypothalamus could serve as the neurochemical basis for control of body temperature. Shortly thereafter, acetylcholine was proposed as an inhibitory neurotransmitter mediating heat gain or heat loss, and peptides were suggested as neuromodulators. Today, a long list of neurotransmitters and neuromodulators (table 3.1) has been reported to change body temperature when injected into the cerebral ventricles or directly into the brain. In addition to this list, recent experiments indicate that the gas NO may also play a role in thermoregulatory pathways.

However, evidence supporting a role for any of these substances' transmitting specific thermal information among neurons in the CNS is equivocal (Blatteis 1981; Myers and Lee 1989). This is so because the vast majority of experiments have been pharmacological in nature. Differences in the methodology employed, species of animals used, route of drug administration, drug dose, anesthetics used, restraint, and ambient temperature have seriously complicated the interpretation of the data collected. Furthermore, in the majority of studies, only a change in body temperature was measured, which could simply be a secondary response to activation of another autonomic physiological system. For serious progress to occur, direct measurements should be made of thermoregulatory effector mechanisms in response to the endogenous release of these putative neurotransmitters. Excellent reviews of this area have been published (Myers 1980; Blatteis 1981; Bruck and Zeisberger 1990; Clark and Fregly 1996).

Since the late 1960s, the most consistent thermoregulatory responses have been elicited by central stimulation with norepinephrine (NE), dopamine (DA), and serotonin (5-HT). However, even these responses vary among species. The anterior hypothalamic-anterior preoptic area is innervated by noradrenergic fibers arising from the ventral noradrenergic bundle, whose cell bodies are located in or near the locus coeruleus. When NE is injected into the ventricles or into specific hypothalamic sites, it produces a fall in body temperature (in cats, rats, and nonhuman primates). These effects are produced through activation of α- but not β-noradrenergic receptors. Furthermore, when core temperature is elevated to about 40°C, NE is released into the hypothalamus of conscious animals (Myers 1980; Quan et al. 1992). This would be an indication of NE activating heat-dissipation pathways in this area of the brain.

These experiments show a correlation between the extracellular concentration of NE and modifications in body temperature, but they do not show that NE is released in the preoptic anterior hypothalamus at specific synapses that participate in the circuitry underlying the control of body temperature. In fact, analyses of the effects of NE on the activity of single neurons in the POAH have shown that NE has an inhibitory effect on thermosensitive as well as thermoinsensitive neurons. In the anterior hypothalamus of the cat, most of the warm-sensitive, cold-sensitive, and thermally insensitive neurons were inhibited by NE and 5-HT (Jell 1974).

Serotonin (5-HT) is the other prominent catecholamine proposed to influence the control of body temperature by the POAH. 5-HT innervation of the hypothalamus comes from fibers arising from cell bodies located in the raphe dorsalis, raphe magnus, and raphe centralis superior, or groups

B7–B9, using the original terminology of Dahlstrom and Fuxe (1964). Many pharmacological experiments have shown that 5-HT injected ICV or into the POAH at certain doses produces hyperthermia. However, at other doses it produces hypothermia. Myers (1980), using push-pull perfusion, found an increase in 5-HT in perfusates from the POAH when the animals were cooled, but not when they were heated.

Again, as for NE, recordings from single-unit activity reported contradictory results: 5-HT seems to produce nonspecific effects (excitation or inhibition of spontaneous activity of thermosensitive or thermoinsensitive neurons). In fact, Watanabe et al. (1986) reported that of the total number of warm-sensitive neurons recorded in the hypothalamus, 91 percent were activated by 5-HT and 85 percent were inhibited by NE. Of the 58 percent that were thermally insensitive, 71 percent were inhibited by NE and 37 percent were excited by 5-HT. In contrast, other studies have suggested that 5-HT increases the inhibitory input from central thermosensors to effector neurons, thus activating the increase of vasomotor tone and thermoregulatory heat production, whereas NE suppresses this inhibitory input. The increased heat conservation and production that may increase body temperature in a thermoneutral environment are the result of NE activation. Increased heat loss and reduced heat production, leading to a decrease in body temperature, are the result of activating the 5-HT system. These changes were observed in the guinea pig and rat (Bruck and Zeisberger 1990).

From the preceding paragraph, it is clear that species differences continue to plague any unified understanding of the role that NE and 5-HT play in thermoregulatory pathways. Moreover, to think that they are the sole mediators of warm and cold information in hypothalamic pathways is unrealistic. That thermal information could reach the hypothalamus after specific spinal fibers reach the midbrain and activate cell bodies of the ascending NE and 5-HT fibers is very likely. However, the role these monoamines play in the anterior hypothalamus-preoptic area remains unclear.

Another prominent amine thought to participate in hypothalamic temperature regulation is dopamine (DA). Lee et al. (1985) reviewed the literature in this field and concluded that in rats, rabbits, and other species (including humans), DA plays a role in the thermoregulatory pathways mediating heat loss. Microinjections of DA into both ventricles and POAH reduce body temperature in the rat (Brown et al. 1982). In the hypothalamus of the cat, Sweatman and Jell (1977) showed that DA microiontophoretically applied to cold-sensitive neurons decreased their spontaneous firing rate. These data indicate that DA exerts its effect by inhibiting cold-sensitive cells, thereby suppressing heat production. This hypothermic effect is further supported by in vitro electrophysiological recordings from tissue obtained from the hypothalamus of the rat. Perfusing hypothalamic slices with DA reduced the activity of cold-sensitive neurons and increased the firing rate of warm-sensitive neurons (Scott and Boulant 1984). These results rule out the possibility that DA depresses all neurons in a nonspecific manner, and also strengthen the hypothesis that DA could be involved in the central control of body temperature by mediating heat dissipation.

DA microinjected into other areas of the brain, particularly in areas involved in motor behavior, also affects temperature regulation. For example, injecting DA into the caudate putamen produces hypothermia that can be antagonized by DA-receptor blockers (Lee et al. 1985). In the substantia nigra, another area of motor control, microinjections of apomorphine (DA-receptor agonist) produced a dose-dependent hypothermia that was antagonized by the central or systemic injection of a specific DA-receptor blocker. This hypothermia was associated with increased heat loss and decreased heat production, and it occurred at different ambient temperatures, thus indicating that apomorphine did not produce a poikilothermic state. Furthermore, the same dose of apomorphine injected into the substantia nigra or the POAH produced similar hypothermic effects (figure 3.8). Moreover, when the substantia nigra was destroyed, resting body temperature increased. However, systemic injections of apomorphine in substantia nigra-lesioned rats and normal rats produced similar hypothermic effects.

Why would dopaminergic areas of the brain such as the caudate putamen, and particularly the substantia nigra, which are involved in motor functions, also participate in thermoregulation? Recall how structures and physiological systems served different functions as organisms changed during evolution (Satinoff 1978). The upright posture of mammals contributed considerably to their increase in endogenous heat pro-

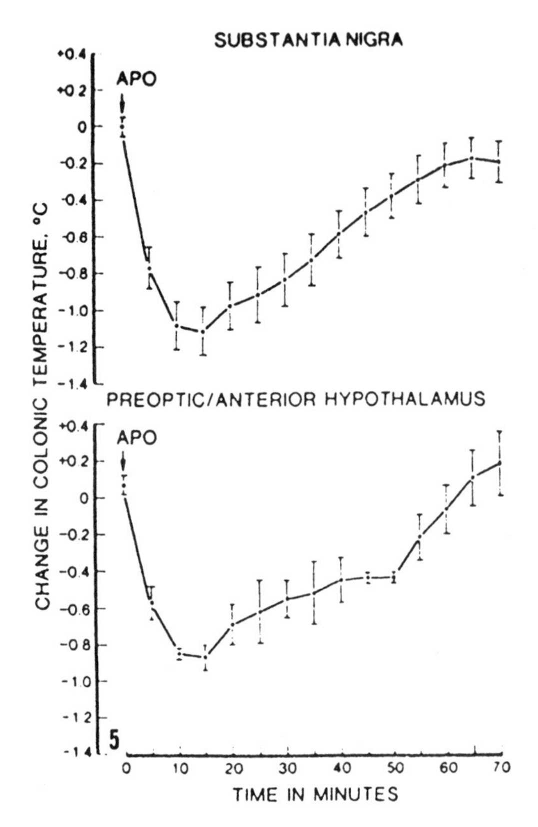

Figure 3.8
Magnitude and time course changes in colonic temperature (mean ± S.E.) produced by bilateral microinjections of 20 µg apomorphine (APO), a dopamine agonist, into the SN and POAH of 5 animals. (From Brown et al. 1982.)

duction (Heath 1968). Could the nigrostriatal pathway have maintained control of both motor behavior and temperature regulation throughout evolution? Patients with Parkinson's disease who have a lesion in this pathway have disturbed motor behavior as well as temperature control. Lee et al. (1985) suggest that the circuitry through which these structures are connected is the anterior hypothalamus via dopaminergic pathways (figure 3.9).

Another group of substances purported to participate in thermoregulatory pathways is the neuropeptides. Numerous papers have been published in this area, but conclusions regarding their role in temperature

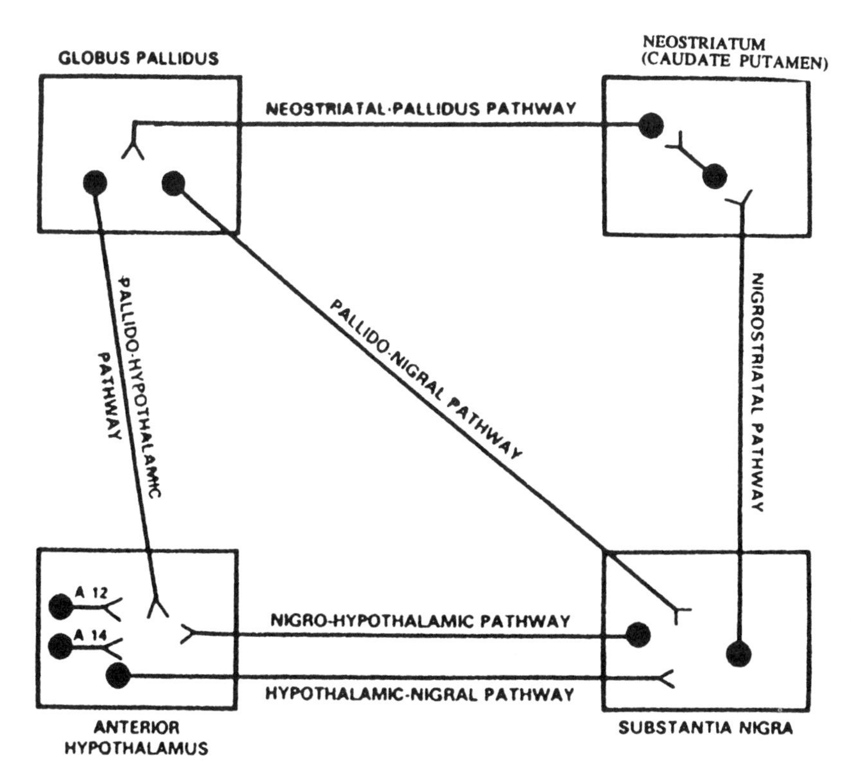

Figure 3.9
Diagrammatic representation of anatomical connections between the anterior hypothalamus and the substantia nigra.

regulation are difficult because of the secondary effects they produce on other physiological systems (particularly the cardiovascular system) when injected centrally. Clark and Fregley (1996) summarized more than 30 peptides known to affect body temperature when injected into the brain by dividing them into three main categories. The first category includes peptides that produce hypothermic effects, such as angiotensin II and neurotensin. In the second category are the peptides that produce hyperthermic effects: (β-endorphins, LH-RH, somatostatin, TRH, and vasoactive intestinal peptide (VIP). The third category includes the antipyretics AVP, (α-MSH, and corticotropin releasing factor (CRF). From critically evaluating the literature on these peptides, Clark and Fregley conclude that (a) there is little evidence to support a role for peptides in the first category as mediators of thermal responses; (b) (β-endorphin is important to the hyperthermic response produced by stress in rats and during hibernation in ground squirrels; and (c) AVP and (α-MSH could be endogenous antipyretics.

Most of the peptides coexist in hypothalamic terminals with monamines and probably amino acids. Thus, dissecting the singular effects of monoamines from those of peptides or amino acids may not elicit a consistent or expected response because under physiological conditions, these substance probably are coreleased and interact within the synaptic cleft to elicit a postsynaptic event.

4

More About the Brain and Temperature

That Round Part of the Brain: The Limbic System

The limbic system has been defined as a conglomerate of functionally related structures that form a ring toward the hilum of each cerebral hemisphere (figure 4.1). It consists of the amygdala, septum, nucleus accumbens, cingulate gyrus, hippocampus, and hypothalamus, as well as related structures. Papez (1937) proposed that neurons in the limbic lobe formed a neuronal circuit that supposedly served as the basis of emotional and cognitive expressions of feelings. McLean (1952) further proposed that the limbic system was important in the expression of emotions, and that the hypothalamus served as the final common output of this system, with the orbitofrontal cortex controlling limbic activity. It is generally accepted that the function of the limbic system is to control and regulate emotional states, motivation (punishment and reward), learning, and memory, as well as metabolic and vegetative functions.

Rewards and pains form the basis of behaviors that subserve the survival of the individual and the species. Feeding, drinking, keeping warm and avoiding extreme temperatures, sexual activity, sleep, and maternal drives are behaviors that tend to satisfy internal needs. These behaviors are coded in the limbic system along with the codes subserving experiences of pleasure and punishment. Feeding and drinking are pleasurable if we are hungry or thirsty, respectively. Sexual activity is pleasurable if we have been deprived of sex. Obtaining warmth in a cold environment is pleasurable. Going to sleep without being threatened by the surroundings or enemies is pleasurable. We tend to perform these behaviors because we are rewarded by them. We are *motivated* to do them because they make

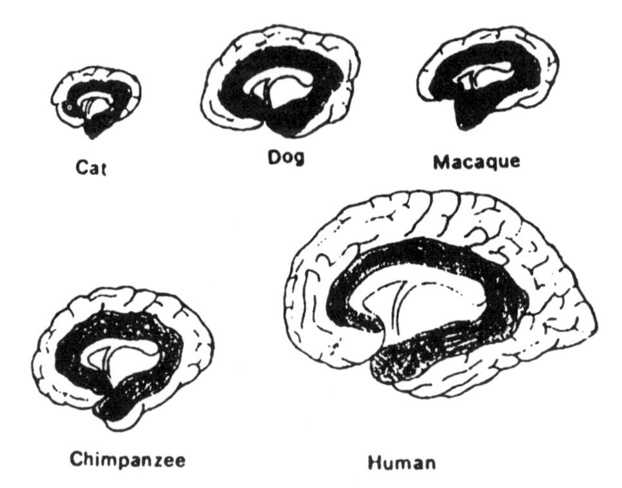

Figure 4.1
Limbic (black areas) systems of the cat, dog, macaque, chimpanzee, and human.

us feel good. That "feeling" is in fact an *emotion*. Life and behavior, as we interpret them today, would have been impossible without motivation and emotion. Both emotion and motivation are, in fact, the guides that orient us in our relationships with the world, including our relationships with other people (Mora 1999).

Pains, Pleasures, and Temperature

What enables an animal to learn to press a bar to obtain heat in a cold environment or activate a fan to keep cool in a hot environment? Probably the same motives that cause the animal to press a bar to obtain food or water if the animal is hungry or thirsty. Animals do those things because they are pleasurable. They are motivated by internal drives to obtain the rewards that are coupled to those behaviors. Heat is reinforcing (rewarding) in a cold environment just as cold is reinforcing in a hot environment. The basis of these behaviors is survival of the individual. Failure to respond leads to punishment.

From lower vertebrates to mammals, behavioral thermoregulation has been the universal strategy used by the brain to maintain its temperature. From the beginning of life, behavioral strategies were used before any autonomic or vegetative functions evolved. These behavioral strategies constitute the basis of survival. In fact, behavioral thermoregulation, as presented in chapter 1, occurs even in unicellular organisms. This capacity was acquired in evolution long before the first nerve tissue appeared.

In mammals, the concept that neural mechanisms for behavioral thermoregulation were rooted in the brain long before autonomic mechanisms has been shown by experiments that lesioned the brain. For instance, lesions of the preoptic anterior hypothalamus (POAH) that impaired autonomic control did not impair the capacity of the animal to perform an operant response (pressing a bar) to obtain heat or cold to maintain body temperature (Carlisle 1969; Satinoff and Rutstein 1971). It is clear, therefore, that behavioral thermoregulation is more primitive than autonomic thermoregulation.

Olds and Milner (1954) demonstrated, for the first time, that electrical stimulation of a specific area of the brain, the septum, was rewarding—the animal learned to press a bar to stimulate its own brain. Soon thereafter, self-stimulation was elicited in all species from fish to mammals, including man (Mora 1977). After that, numerous experiments uncovered different areas of the brain that supported the phenomena of intracranial self-stimulation. Rolls and colleagues (1980) described the neurophysiological circuitry involved in self-stimulation of the monkey. Neurons in different areas of the limbic system—the hypothalamus, amygadala, and prefrontal cortex, among others—were shown to be interconnected because self-stimulation of one of these areas activated neurons in another. Moreover, the biological significance of self-stimulation in the hypothalamus was revealed. The same neurons that activated self-stimulation were activated by the sight and taste of food when the animal was hungry, that is, when the food was rewarding (Rolls et al. 1976). A later study (Mora and Cobo 1990) showed that self-stimulation of just one area evoked activity in many, and perhaps redundant, circuits of the brain (figure 4.2).

In any living creature, from fish to human, it is obvious that heat or cold can be rewarding, depending on environmental temperature. Fish living in a determined ambient temperature, when placed in a colder or warmer environment, move to the temperature to which they were acclimated. Rats placed in a cold chamber will press a bar to turn on a heat

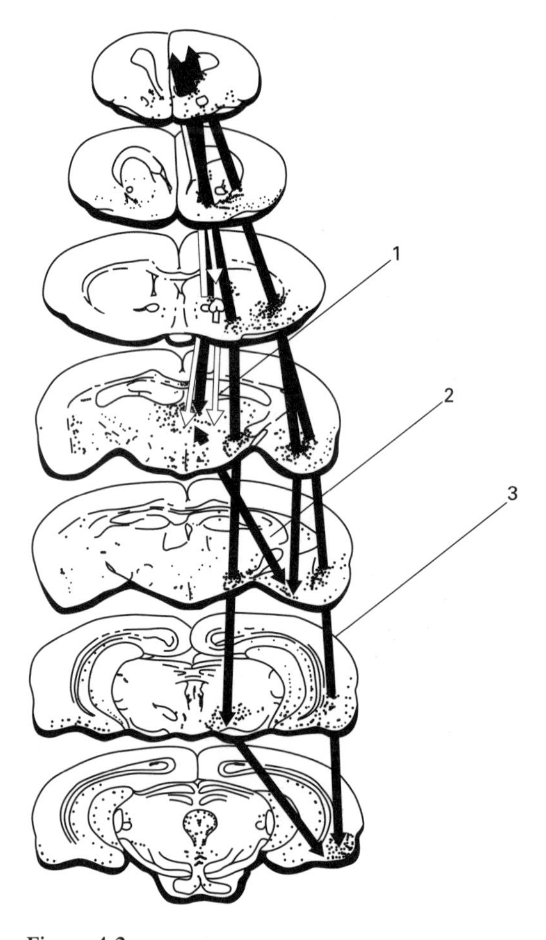

Figure 4.2
Some pathways presumably involved in reward produced by self-stimulation of the prefrontal cortex (limbic area). The circuits depicted are (1) prefrontal-mediodorsal thalamic nucleus-neostriatum-prefrontal; (2) prefrontal-mediodorsal thalamic nucleus-amygdala-prefrontal; (3) prefrontal-entorhinal cortex-ventral tegmental area-prefrontal. (Taken from Mora and Cobo 1990.)

lamp (Weiss and Laties 1961). Rats placed in a warm environment learn to press a bar for cold air. Under these conditions, heat or cold will be rewarding, if the animal works to obtain the desired temperature.

Supposedly the brain circuits for reward described above are part of the brain circuitry that participates in the specific rewards provided by cold or heat when the animal is placed in a hot or cold environment. It is not known if there are specific circuits for specific rewards or simply a unique circuit recognizing reward and other circuits, with further specificity, for the different rewards that subserve different motivated behaviors.

The POAH is at least one brain area that integrates autonomic and behavioral thermoregulation (Satinoff 1964; Corbit 1969). Satinoff (1964) was one of the first investigators to show that a rat can be motivated to work by pressing a bar to provide heat (turning on an infrared lamp) when the POAH is cooled directly through a thermode implanted in it (figure 4.3). Rats also will turn on a draft of cool air when hypothalamic temperature is elevated (Corbit 1969). Most interesting is that rats will work to produce changes in their own hypothalamic temperatures when either skin or hypothalamic temperature is increased above neutrality. In this case, by pressing a bar, rats changed the temperature of the water perfusing a hypothalamic thermode, causing an abrupt change in skin temperature and hypothalamic temperature (Corbit 1969). This last study supports previous findings, and further demonstrates that discomfort resulting from deviations in skin or hypothalamic temperature can be alleviated by changing hypothalamic temperature.

Unfortunately, experiments have not been done to demonstrate the reward areas of the limbic system most relevant to behavioral thermoregulation. Three areas of the limbic system—the lateral hypothalamic-medial forebrain bundle, the nucleus accumbens, and the septum—are tightly interconnected with the POAH, and are involved in both reward and temperature regulation (Boulant 1980; Dean and Boulant 1989; Wang et al. 1994). Moreover, the hippocampus, another area of the limbic system involved in learning and memory, has been shown to send important afferent connections to thermosensitive neurons (type "b" of Boulant; see Chapter 3) in the preoptic area. Boulant (1996) has suggested that these neurons may participate in behavioral thermoregulation.

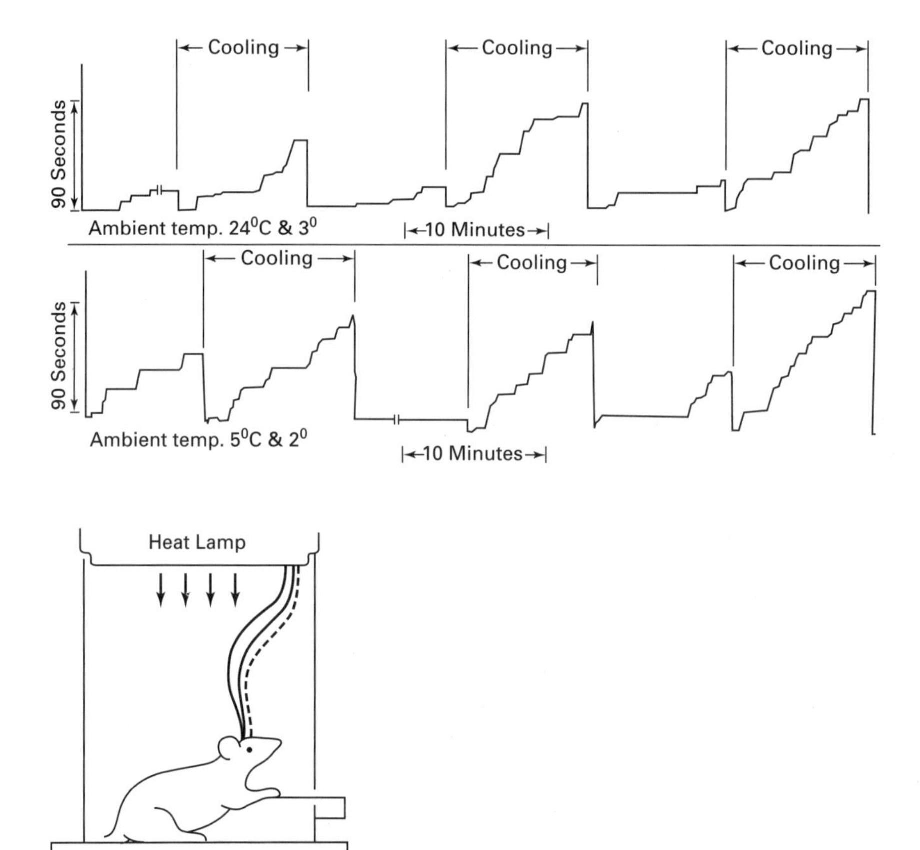

Figure 4.3
Schematic of a rat bar-pressing to turn on a heat lamp in response to cooling the POAH. Also shown is the cumulative record of the amount of bar-pressing before and during (labeled cooling) hypothalamic cooling in a neutral (24 ± 3°C) and cold (5 ± 2°C) environment. The amount of heat the animal worked for was greater in the cold environment. (Modified from Satinoff 1964.)

Circadian Rhythms: The Case for Thermoregulation

Almost everything in the universe is cycled. All living organisms, from unicellular animals to humans, have periods of high and low activity. Many physiological functions also have periods or cycles, from the most obvious, sleep-wakefulness, to feeding, drinking, and so on. Most of these periods or rhythms have the same cycle as the rotation of Earth and its major consequence, the light-dark period. This period of light and darkness is called circadian because the daily rhythm usually is not exactly 24 hours. The Latin word *circa* means "near," and *diem* means "day" (Aschoff 1965; Rusack 1989; Heller et al. 1990).

Apart from the obvious circadian rhythms of behavior, hormones, sleep-wakefulness, and so on, mammals have a circadian cycle of body temperature that is independent of the sleep cycle. In humans, this body temperature cycle persists even during sleep deprivation (Kreider 1961; Kleitman 1963). It also is independent of ambient temperature and persists until body temperature falls below 6°–16°C, depending upon the species studied (Folk 1974). Figure 4.4 shows simultaneously recorded rhythms for activity, heart rate, body temperature, and blood pressure over 5 consecutive days in a white rat. The night cycles are best identified by the troughs in the heart rate data. Although the 4 cycles appear to be somewhat synchronous, statistical analysis shows that they are indeed separate. Heart rate, blood pressure, and activity cycles peaked before the body temperature cycle. These data provide evidence that the cycle of body temperature is not responsible for the heart rate and blood pressure rhythms, previously believed to be true for humans.

In a series of human experiments performed by Wenger and colleagues (1976), the sweating threshold during physical exercise was lower at night than during the day, suggesting that humans thermoregulate during the night at a different level than during the day. That the circadian rhythm for body temperature is deeply rooted in the brain and independent of the sleep-wake, drinking, feeding, and motor activity cycles was shown by Satinoff (Satinoff and Prosser 1988). She lesioned the suprachiasmatic nucleus and found that the rhythms for drinking and motor activity were abolished, but the rhythm for body temperature remained.

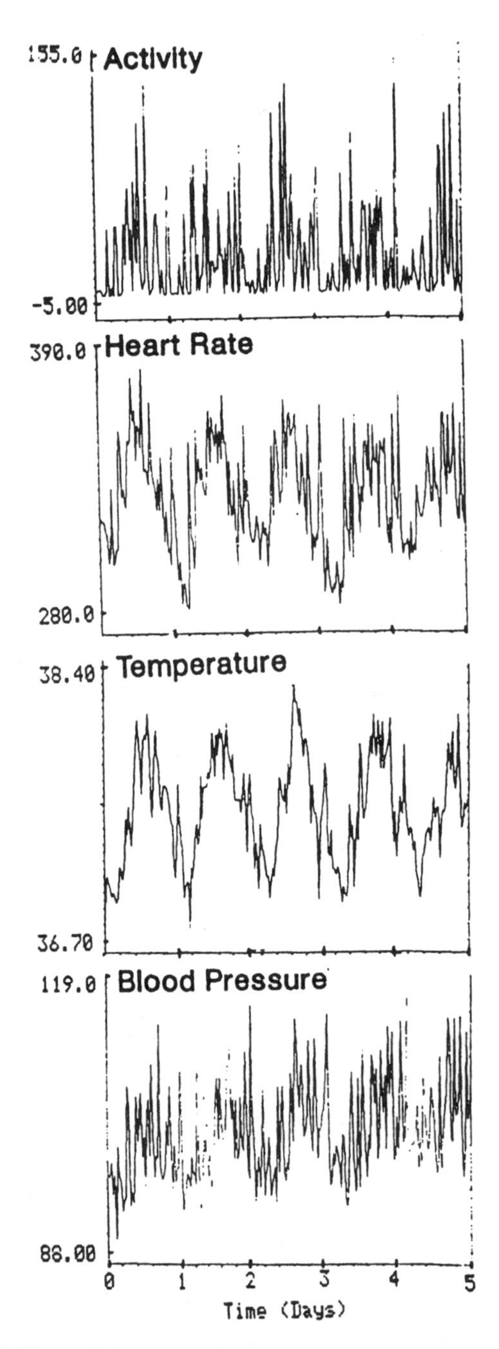

Figure 4.4
Activity, heart rate, core temperature, and blood pressure responses simultaneously transmitted by an implanted radio capsule in a white rat over a period of 5 days. Data were recorded every minute. (Taken from Folk et al. 1995.)

Is Body Temperature Reptilian During Sleep?

Circadian rhythmicity is best represented by the sleep-wake cycle. It has been demonstrated that in birds and mammals a strong interaction exists between sleep and body temperature (Parmegiani 1980; Heller et al. 1990). Cycle variations occur in body temperature during the day and during the menstrual cycle, as they do in many systems. Sleep is a physiological state to which we normally dedicate one-third of our lives, yet the mechanisms responsible for this behavior remain unknown. Nor do we know the biological significance of sleep.

The regulation of body temperature is a discontinuous process during sleep; that is, it is controlled in some periods of sleep but not in others. In other words, during certain periods of sleep we become poikilothermic (like fish or reptiles), whereas during other periods we are homeothermic. This dichotomy presents important implications for understanding the central concept of homeostasis, and above all, it shows the deep interaction between thermoregulation and sleep. To understand the relationship between temperature and sleep, two phases of sleep will be considered: (a) synchronized sleep, slow-wave sleep (SWS), or nonrapid-eye-movement (NREM) sleep, and (b) desynchronized sleep, paradoxical sleep, or rapid-eye-movement (REM) sleep.

A universal phenomenon is feeling cold at the onset of sleep. In fact, at the beginning of NREM sleep, when ambient temperature is within the thermoneutral zone, a fall in hypothalamic and body temperature occurs in numerous mammalian species, including rats, cats, and humans. In humans this brain cooling process has been shown to be independent of the circadian variation of temperature (Obal 1984) and of the reduction in motor activity during sleep (Barrett et al. 1987).

The decline in brain temperature during the initial period of sleep has been attributed to a decrease in sympathetic activity and, as a consequence, a decrease in metabolic rate. During the initial phase of sleep, a change in set-point could explain the change in hypothalamic temperature. In humans, changes in the sweating response are consistent with the findings that during NREM sleep the individual thermoregulates, but becomes poikilothermic during REM sleep. Thus, when ambient temperature is increased during NREM sleep, sweat is produced in proportion to the thermal load, whereas during REM sleep, sweating is absent (Henane et al. 1977). In many mammalian species, the transition from NREM sleep to paradoxical sleep is accompanied by a rise in hypothalamic temperature that is independent of ambient temperature. This has been attributed to various mechanisms, including an increase in brain metabolic rate and a rise in brain blood flow. This phenomenon follows the suspension of thermoregulatory vasomotion in the skin, mucosa, and ear pinna.

During NREM sleep, cats exposed to a cold environment respond by shivering, but as soon as the cat enters REM sleep, shivering ceases. Similarly, cats exposed to a high ambient temperature during NREM sleep become tachypneic, but when they enter REM sleep, respiratory rate decreases. This cessation of autonomic thermoregulatory function is maintained for the entire period of REM sleep (Parmegiani and Rabini 1967, 1970). Cooling or warming the hypothalamus, instead of exposing the animal to a cold or hot environment, produces similar responses. These observations show that during REM sleep, autonomic or vegetative regulation of body temperature is not evoked in response to thermal challenge. According to Parmegiani (1980), the phenomenology of REM sleep is not simply the result of a change in threshold or gain of the different thermoregulatory mechanisms, but an inactivation of hypothalamic thermoregulatory mechanisms resulting in a true poikilothermic condition of the organism. It is an inactivation of the hypothalamic centers controlling spinal cord (shivering and vasomotor) and brain stem (tachypnea) effector mechanisms. He further suggests that the poikilothermic state during REM sleep is produced by an inactivation of the hypothalamic thermostat.

This interpretation is consistent with the model proposed in chapter 3 of a hierarchical control of temperature through different levels of the neuraxis. Thus, during REM sleep, animals and humans thermoregulate, but over a wider core temperature range, and the thermoregulatory effector mechanisms are controlled by thermostats other than the hypothalamus (Horne 1988). The biological significance of this change in the level of thermoregulation during REM sleep remains unclear. A similar physiological example in which hypothalamic control of thermoregulation is acquired progressively with time is that of the newborn infant.

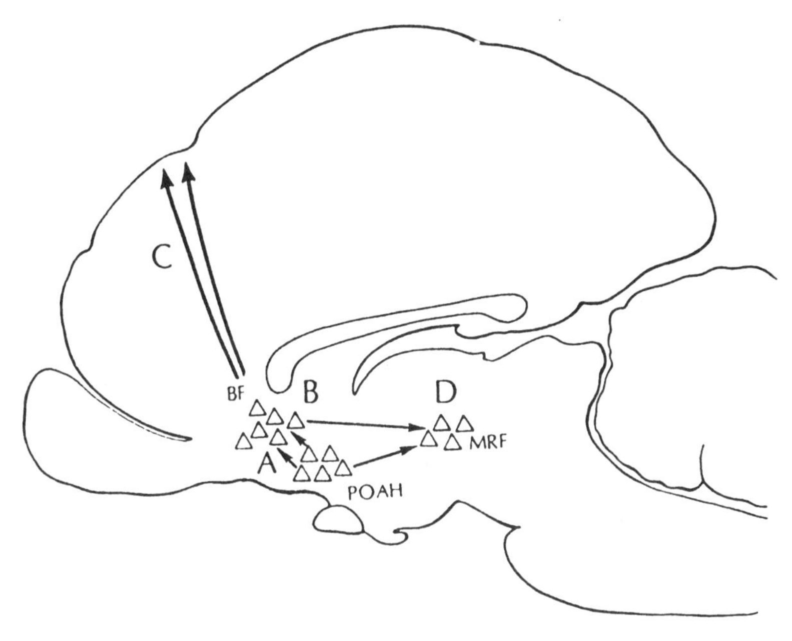

Figure 4.5
Schematic diagram showing the neurobiological circuitry through which heating brain and body induces NREM sleep. Increased body and brain temperature will increase the firing rate of warm-sensitive neurons in the preoptic anterior hypothalamus, which in turn will activate neurons in the basal forebrain. These basal forebrain neurons will produce an inhibitory input to neurons in the cerebral cortex and the medial reticular formation, giving rise to NREM sleep. C, cortex; BF, basal forebrain; POAH, preoptic anterior hypothalamus; MRF, medial reticular formation. (Taken from McGinty and Szymusiak 1990.)

Keeping a Cool Brain: A Function of Sleep?

Does sleep function, in part, to keep the brain cool, or is it just a biological consequence of the cyclicity that all events in nature have? An intriguing hypothesis suggests that NREM sleep is biologically necessary because it provides a homeostatic mechanism for brain and body cooling (McGinty and Szymusiak 1990). NREM sleep seems to represent an integrated thermoregulatory process because it evokes a decrease in metabolism and activates heat loss by increasing skin blood flow and by initiating sweating (figure 4.5). Evidence that these responses are controlled by changes in hypothalamic thermosensitivity is supported by the finding that, compared with being awake, NREM sleep decreases the slope of the metabolic response to hypothalamic cooling. It is fascinating that in dolphins, which basically have two brains and show unihemispheric sleep, the sleeping brain cools (Kolvalzon and Mukhametov 1982).

McGinty and Szymusiak (1990) further suggest that an elevated waking temperature in mammals and birds could not be sustained without the brain cooling provided by sleep. In humans, raising brain temperature by vigorous exercise during the day increases NREM sleep at night; however, performing the same exercise but reducing the rise in brain temperature by cooling the body prevents the increase in NREM sleep. These observations led to the speculation that the brain somehow stores information about how long brain temperature is elevated during the day and then compensates for this "awake heat load" by sleep-related cooling (McGinty and Szymusiak 1990). This speculation is supported by the fact that sleep deprivation increases NREM sleep. Moreover, sleep-deprived humans lower their body temperature during the day (Froberg 1977). These interesting suggestions nonetheless are controversial (Almirall et al. 1993).

Thermoregulation in the Newborn Infant

Nurses who work in intensive care units are keenly aware of the labile nature of body temperature in newborn infants. Newborns (particularly those with low birth weight) do not have mature neural structures controlling endothermy.

Compared with adults, newborns have a high surface-to-mass ratio, which enhances convective, evaporative, and radiative heat loss to the environment. Their tissue insulation also is less than adults' (Hey and Katz 1970), and they have a low metabolic rate (when normalized for surface area). As a consequence, the newborn can rapidly become hypothermic in a cold environment.

During the period of time that newborns are acquiring autonomic control of body temperature, behavior seems to play an important role in thermoregulation. This is reminiscent of the trend in evolution that behavioral thermoregulation preceded the development of autonomic control mechanisms (ontogeny versus phylogeny). During exposure to cold or warm environments, newborns increase or decrease their spontaneous limb movements, respectively. As a consequence, metabolic heat production rises or falls significantly. Moreover, during cold exposure, nonshivering thermogenesis is an important factor in newborns compared with adults, because newborns do not shiver. Shivering occurs only when (a) nonshivering thermogenesis has reached a maximum, (b) body temperature has decreased markedly (Darnal 1987), or (c) the child is exposed to extreme cold (Adamson et al. 1965). Brown adipose tissue has been related to this phenomenon. Brown fat is distributed differently in newborns and adults. In the newborn, this tissue is concentrated in masses, whereas in adults it is dispersed in muscles. During growth and development, nonshivering thermogenesis gradually diminishes (within a few weeks), and there is a corresponding increase in shivering capacity and regression of brown adipose tissue.

There also are contrasting changes in thermoregulation between newborns and adults during sleep, particularly REM sleep. In newborns and infants, the duration of REM sleep is longer than in adults. This has been attributed to the role of REM sleep in CNS maturation (Tolaas 1978). It appears that newborns have full control of body temperature during REM sleep, in both cold and warm environments, which contrasts with the poikilothermic state of adults during this period of sleep. In an exhaustive review of the literature, Bach et al. (1996 p. 392) summarized these contrasting phenomena in the following way:

> The model of inactivation of hypothalamic control and consequently the depression of thermoregulation during REM sleep, cannot be applied to newborns. In fact, thermoregulatory responses to thermal loads are fully efficient and stable in the narrow range of air temperatures usually studied. This suggests two alternative hypotheses: either the functional control of diencephalic structures may not be suppressed during the REM sleep or the autonomy of the rhombencephalon is capable of thermoregulatory control within the narrow range of ambient temperatures tolerated by the newborn.

Heat Loss Overdrive: Hot Flashes

> My husband now sleeps in the living room, but at least I have my hot flashes to keep me warm. (Kronenberg 1990 p. 52)

Why discuss hot flashes in connection with circadian rhythms and sleep? Although hot flashes have been described as occurring sporadically, erratically, and unpredictably, the limited evidence available indi-

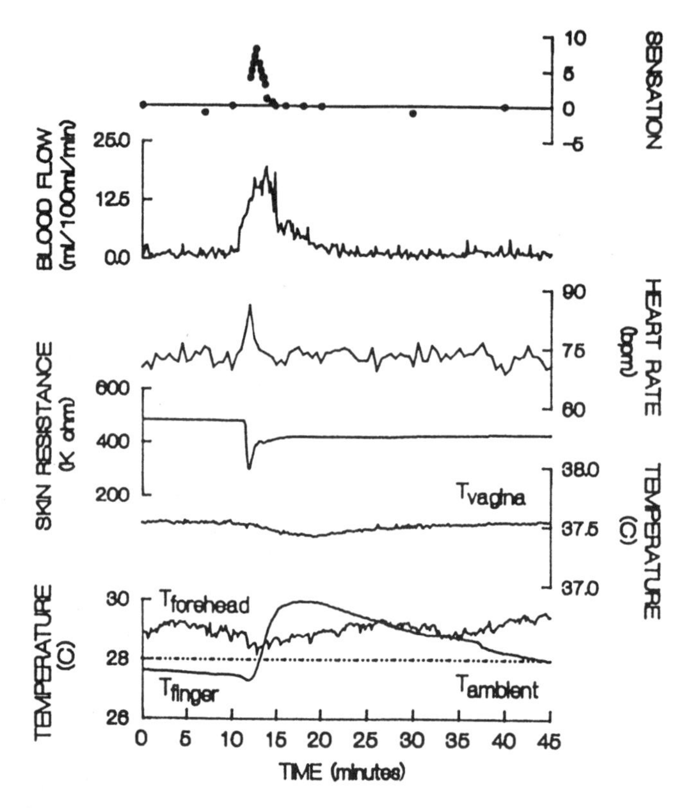

Figure 4.6
Physiological and perceived changes during a typical hot flash in a neutral environment (28°C). Sensation, finger blood flow, heart rate, skin resistance on the chest, vaginal temperature, and skin temperatures on the forehead and finger are shown. (Taken from Kronenberg 1990.)

cates that they are not random, but occur with some regularity (Kronenberg 1990). In fact, they may have a circadian or ultradian periodicity. They usually are more intense at night than during the day, and are more frequent in a warm environment than under cool ambient conditions. In the review by Kronenberg (1990, p. 52), a women commented:

> I turned down a proposal for marriage. He lives in Alabama. If it had been Alaska, I would have said yes.

The sequence of physiological changes that characterize the hot flash is depicted in figure 4.6. Within 60 seconds of the start of a hot flash, the sensation of "hotness" is experienced despite no change in body temperature. During this period heart rate and skin blood flow begin to rise. With the onset of the hot flash, skin resistance drops sharply, skin blood flow rises as much as 30-fold, heart rate continues to increase 8–16 beats/min, finger temperature may rise 7°C, and sweating commences (Kronenberg 1990). Although these physiological changes are perceived primarily in the upper body, similar changes occur in the legs and feet, which indicates that the hot flash is a generalized physiological phenomenon.

The mechanism(s) responsible for hot flashes are not known. They have been associated with low plasma estrogen levels because they typically begin with the onset of menopause (a condition characterized by low estrogen levels), and are alleviated by estrogen replacement therapy. However, there are numerous exceptions to this generalization. On the other hand, the fact that estrogen excites warm-sensitive neurons and inhibits cold-sensitive neurons in the POAH, and is known to enhance the vasoconstrictor response to vasoactive drugs, argues for both a central and a peripheral role for estrogen in the mechanism of hot flashes.

The premonition that a hot flash is about to occur may be explained by a sudden decline in set-point (figure 4.7). This is just the opposite of what occurs with fever (see chapter 8), that is, the set-point increases. During a hot flash, the error signal produced by the difference between core temperature and set-point may account for the sensation of hotness, whereas the difference between core temperature and set-point in fever is the reason for feeling cold. Activation of heat loss mechanisms (increased skin blood flow, sweating) is driven by the body's desire to achieve the new lower set-point temperature during a hot flash. The fall in core body temperature, as illustrated by the decline in vaginal temperature in figure 4.6, is the key observation.

If the activation of heat loss mechanisms were nonspecific, thermoregulatory responses would be activated to counteract the fall in core temperature (Lomax et al. 1980). Contrary to what occurs during fever, the typical hot flash lasts only for minutes and a new thermal equilibrium at the lower set-point is never achieved. With fever, the new elevated set-

Fever	Hot flash
1. **Onset of heat production and heat conservation mechanisms (vasoconstriction, shivering) before any change in T_b**	1. **Onset of heat loss mechanisms (vasodilation, sweating) without any change in T_b**
2. Behavior promoting heat conservation	2. Behavior promoting heat loss
3. $\Uparrow T_b$	3. $\Downarrow T_b$
4. Onset of heat loss mechanisms (vasodilation, sweating)	4. Onset of heat conservation and heat production mechanisms (vasoconstriction, shivering)
5. Behavior promoting heat loss	5. Behavior promoting heat conservation
6. $\Downarrow T_b$ to normal	6. $\Uparrow T_b$ to normal

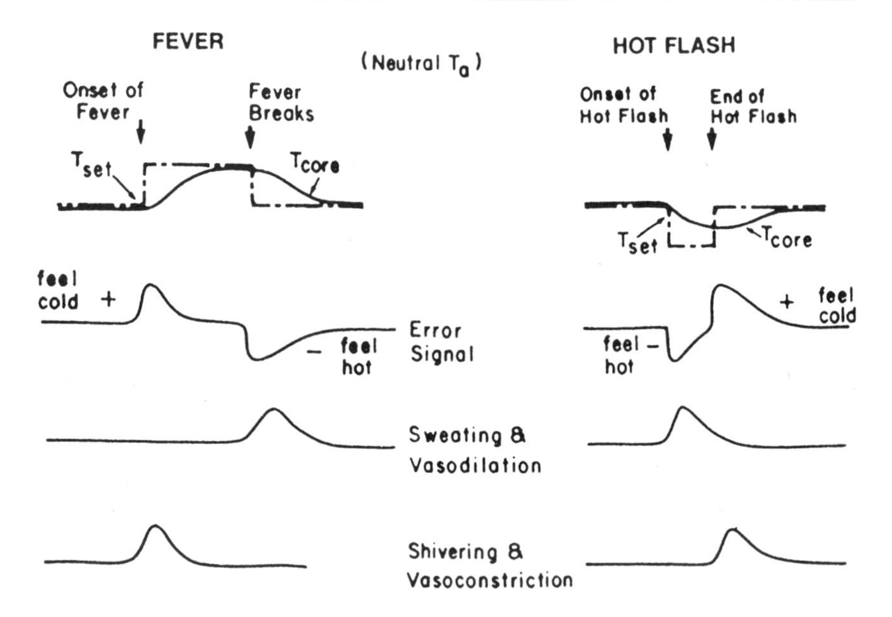

Figure 4.7
Schematic diagram comparing the effects of fever and a hot flash on set-point, core temperature, effector mechanisms, and sensation. (Modified from Kronenberg and Downey 1987.)

point temperature is achieved and maintained for hours or days. We have learned much about fever over the last several decades because of numerous animal models available. Unfortunately, there are no animal models available to study hot flashes, although the adult female monkey does show some cycling in skin temperature similar to what occurs during a hot flash.

5

What's So Important About a Body Temperature of 37°C?

The Temperature Straitjacket

All animals and humans are constrained by the environment in which they live, especially the temperature of their surroundings. The temperature extremes on Earth range from –68°C (–128°F) in cold surroundings to 88°C (190°F) in desertlike environments, but the range of ambient temperature that allows for animal life is only about 0° to 50°C (the *temperature straitjacket*). Most animals live within a more limited range (Prosser 1961). The average temperature on Earth is about 16°C (61°F), but if the coldest months of the year are excluded, it is approximately 23°C (73°F).

Interestingly, the earliest civilizations on Earth were located in areas where environmental temperature ranged from 21° to 26°C—North Africa, Persia, northern India, southern China, Peru, Chile, Argentina, southern Brazil, southern Central Africa, and central Australia (Prosser 1961). This average temperature becomes important when trying to predict the body or brain temperature that humans "regulate." The different mechanisms for surviving the "straitjacket" are varied and intriguing. For example, some animals "supercool" or actually allow part of their body water to freeze. Other animals hibernate or become dormant.

Throughout evolution, multicellular organisms had to cope with a constantly changing environment within the temperature straitjacket. During those early times, and for periods lasting more than a billion years, organisms were enslaved by the temperature of their surroundings. These organisms, which included fish, amphibians, and reptiles, were cold-blooded and were classified as "ectotherms" because they relied on external sources of heat (see chapter 2). When ambient temperature rose, these animals were active and sought food and water. When ambient temperature declined, their motor activity decreased until they became dormant.

As we have seen in previous chapters on the evolution of the brain, during the transition from reptiles to mammals, mammals evolved a body in parallel with a brain capable of controlling its own body temperature. In contrast to ectotherms, mammals and birds rely on internal heat production to regulate their temperature and are therefore "endotherms." This was a revolutionary development because it made it possible for these animals to explore outside their restricted ecological niches. Mammals were equipped with this capacity to maintain a constant body temperature through an exquisite balance between heat production and heat loss.

Mammals are inefficient metabolic machines. In fact, more than 70 percent of their energy intake in the form of fuel is transformed into heat. Under basal conditions the tonic contraction of muscles provides about 30 percent of the total heat production. When active, muscles are powerful heat pumps (through contraction) producing much more heat than is needed to maintain body temperature. In addition to this heat energy from metabolic processes, heat can be gained from the environment, especially radiant heat from the sun. This considerable amount of heat produced or gained by our body, in excess of that required to maintain body temperature constant, is eliminated through a sophisticated system controlled by the brain. The avenues of this heat loss are radiation, convection, conduction, and evaporation (Slonim 1974) (figure 5.1).

Heat transferred by *radiation* is electromagnetic waves from surrounding objects (ceilings, walls, floors, other people, furniture, etc.). If these objects are warmer than our skin (about 93°F), heat is gained by radiation. If they are colder than the skin, heat is lost by radiation. Radiant heat from the sun is always a concern and can add substantially to the thermal loan on an organism. Another concern regarding radiant heat from the sun is the possibility of contracting skin cancer (melanoma). This is one of the most virulent forms of cancer and warrants the use of sunscreens to protect the skin.

Heat loss by *conduction* is the transfer of heat in a solid, liquid, or gas from one molecule to another; it depends upon the conductivity of the

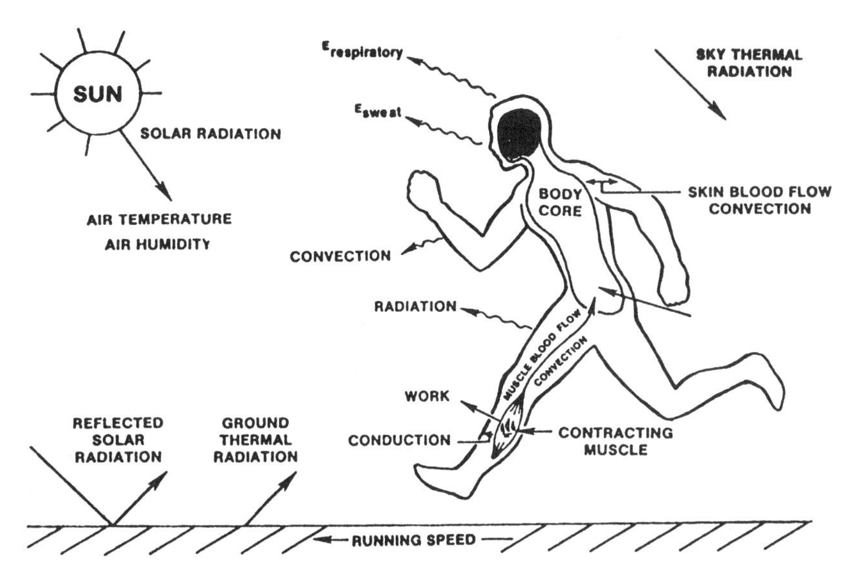

Figure 5.1
Schematic diagram showing human heat production, transfer, and loss. Heat produced within working muscles is carried by the blood to the body core (abdominal, thoracic, cranial cavities), thus elevating core temperature. This rise in core temperature results in the transfer of blood from the core to the skin and the initiation of sweating. Heat is lost from the surface of the skin by radiation, convection, conduction, and evaporation. (From Gisolfi and Wenger 1984.)

substance and the temperature difference between the points in question. For example, a cold steel ball held in a warm hand slowly warms because heat from the skin causes the molecules of steel directly in contact with the skin to vibrate. This vibration increases the temperature of those molecules, which, in turn, will cause the molecules next to them to vibrate. Thus, there is no transfer of matter during conduction. A warm steel molecule does not migrate from one side of the steel ball to the other. This is different from convection, where there is a transfer of matter. Conduction is not a major avenue of heat loss for humans because they seldom come in contact with objects colder than their skin.

A major avenue of heat loss for humans is *convection,* which occurs in gases and liquids via a mixing process. For example, if a subject slides into a pool of cold water and remains absolutely still, heat is lost primarily by conduction. Cold water molecules are heated by warm skin, and after a short period of time, the subject feels less cold because the layer of the water directly in contact with the skin has been warmed. However, if the surrounding water is circulated, or if the subject begins to swim, a much colder feeling ensues because cold water molecules continually come in contact with warm skin, become heated, and move away, taking heat with them. There is a transfer of matter because, in our example, cold water molecules come in contact with warm skin and move away. If the water is warmer than the skin and it is circulating, heat will be gained by convection.

Evaporation is the process of transforming a liquid into a vapor without changing its temperature. Approximately 25 percent of our heat loss, even in a cool environment, is by evaporation. This process is called *insensible perspiration* and does not require sweat glands. Fluid evaporates continually through the pores in the skin.

The Thermoregulatory Computer

What Is Regulated?

The term "body temperature" is a misconception. When humans are placed in an environmental chamber and ambient temperature is slowly raised from a level that is cold to one that is hot, skin temperature changes with the environment, but the temperature of other parts of the body, such as the heart, remain constant (figure 5.2). It is as if we had, from a thermoregulatory viewpoint, two bodies in just one organism, that is, the reptilian or poikilothermic body as a shell and a homeothermic core. In humans, the core consists of the abdominal, thoracic, and cranial cavities; the shell comprises the skin, subcutaneous tissue, and muscles. Therefore, what we typically refer to as body temperature (T_b) is a composite of the core (T_c) and the shell (T_s) temperatures:

$$T_b = 0.65\ T_c + 0.35\ T_s.$$

The numbers in the above equation represent the magnitudes of these two bodies in a neutral environment. In fact, these two bodies are not fixed anatomical entities because they change with ambient temperature, and therefore they can get larger or smaller. Thus, the concept of core and shell is more physiological than anatomical. In a sense, anatomy and

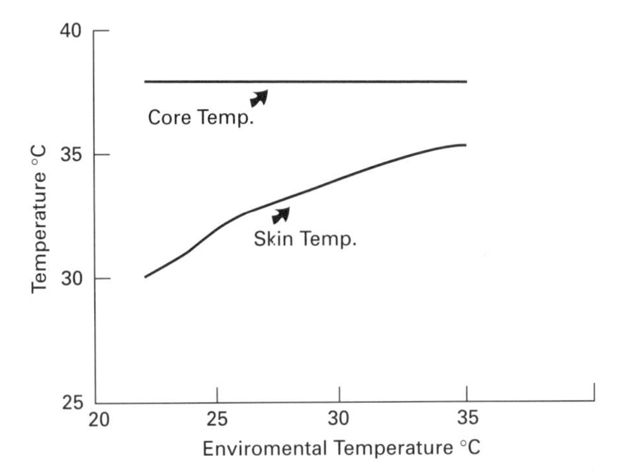

Figure 5.2
Graph of core and skin temperatures as ambient temperature is increased from cool (22°C) to hot (35°C). Even though the core-to-skin temperature gradient decreases as ambient temperature increases, heat transfer from core to skin is maintained by an increasing skin blood flow as ambient temperature rises. Heat loss is maintained over a wide ambient temperature range because an increase in evaporative heat loss as ambient temperature rises compensates for the decrease in radiative and convective heat loss.

physiology are divorced. In this anatomical-physiological interplay, blood flow to the skin plays a crucial role. By changing cutaneous blood flow and cutaneous blood volume, the insulative value of the shell can be reduced or increased.

In a cold environment, skin blood flow and blood volume decrease, which increases the insulative value of the skin. Thus, the skin assumes a greater role in determining body temperature in a cool environment, as reflected by the change in coefficients in the equation for body temperature:

$$T_b = 0.6T_c + 0.4T_s.$$

In a warm environment, blood flow is diverted to the skin to help transfer heat from the core. As a result, the insulative value of the shell decreases. Thus body temperature is defined as

$$T_b = 0.8T_c + 0.2T_s.$$

The most important difference between core and shell temperatures is that core temperature is regulated, whereas the shell temperature is influenced by environmental temperature and skin blood flow. In a sense, the shell serves at the pleasure of the core. It can be enlarged or contracted by manipulating skin blood flow to facilitate heat loss or heat gain, respectively.

Shell temperature is frequently measured by placing temperature-measuring devices on numerous skin locations and averaging the values. However, this mean skin temperature is potentially misleading if each individual temperature is not weighted according to the *area* and *sensitivity* (temperature-receptor density) of the site represented. Core temperature is assessed by measuring rectal, armpit, ear canal, tympanic membrane, central blood (right heart), oral, hypothalamic, or esophageal temperature. In many hospitals, ear canal temperature is taken as a measure of core temperature as a matter of convenience; however, it is highly variable because it is influenced by changes in ambient temperature.

The Central Computer

The system that allows us to regulate body and brain temperature around 37°C is exquisite in design and execution. (See table 3.2 for a description of the neurotransmitters and neurochemical pathways found in the hypothalamus and the interaction of these systems through synaptic as well as volumetric forms of neural communication.) In fact, it is one of the best examples of physiological regulation in the human body. It consists of temperature sensors located throughout the body (including the brain), a complex neural network located primarily in the anterior preoptic hypothalamus and limbic system that receives and integrates this thermal information, and specific neuronal populations controlling behavioral and physiological responses. In addition to thermal information, the neurons in these integrative circuits receive nonthermal input that can influence body temperature. This nonthermal input consists of endogenous substances like interleukin–1 and prostaglandin E that can produce fevers, as well as hormones including testosterone, estrogen, and progesterone (see chapter 8). Temperature-sensitive neurons also respond to changes in osmolality, and to glucose and ion concentrations in the soup that bathes them.

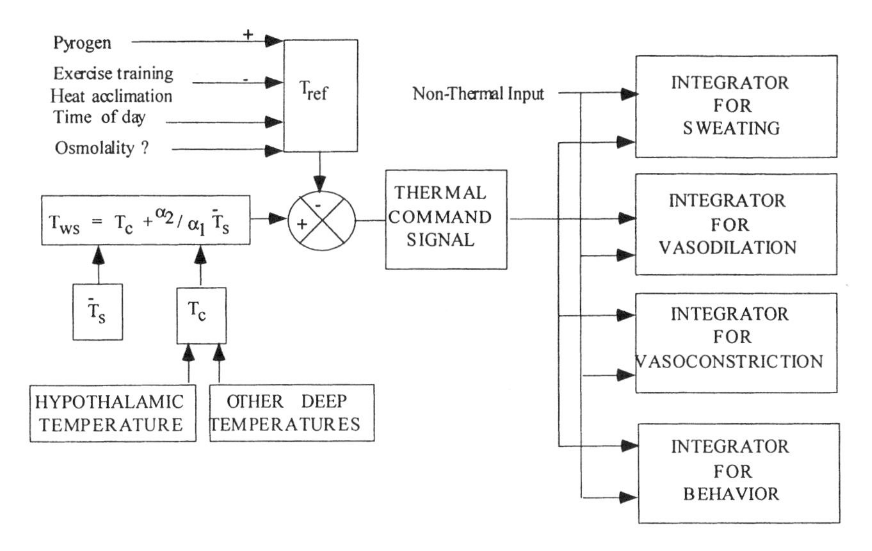

Figure 5.3
Schematic diagram illustrating the elements that control the temperature regulatory effector responses. Basic thermal input is transmitted from skin (T_S) and core (T_C) thermoreceptors. This information is integrated and a "weighted sum" (T_{WS}) of this activity is then compared with basal neural activity (T_{ref}) within the regulatory circuiting to form a "command signal" that impinges on all integrators for regulating brain temperature. The ratio of $\alpha 1/\alpha 2$ is 9/1. (From Gisolfi and Wenger 1984.)

These elements of the control system are depicted in figure 5.3. Each box in the diagram consists of multiple neuronal circuits that receive excitatory and inhibitory inputs resulting from the release of chemical substances called neurotransmitters. (A hypothesis explaining the acquisition of this machinery through the evolution of the brain in mammals is developed in chapters 2 and 3.) As illustrated in the diagram, the integrative circuits compare the thermal and nonthermal input that they receive with basal neuronal activity, then generate what is described as a "command signal." This command signal, in turn, serves as the input to separate populations of neurons controlling sweat rate, shivering, blood flow to the skin, and behavior. In addition to this command signal, these separate neuronal pools subserving the thermoregulatory effectors receive nonthermal input that can influence their ultimate responsiveness.

The operation of the temperature regulatory system is no doubt a complicated function of thermal and nonthermal inputs. However, it is instructive to consider its operation as simply a linear function of thermal input from the core and the shell. The equation below describes the control of the different thermoregulatory effector responses in this fashion (Gisolfi and Wenger 1984). The R represents any effector response, R_0 is a basal value of R, α_1 and α_2 are coefficients, and b is a constant term:

$$R - R_0 = \alpha_1 T_c + \alpha_2 T_s - b.$$

In humans, the ratio of α_1/α_2 is 9:1 for all heat-dissipating responses. These include the sweating response and alterations in blood flow to the skin. However, in a cold environment, heat gain responses are influenced more by mean skin temperature. In both men and women, skin temperature contributes about 20 percent of the control of shivering and vasoconstriction (Cheng et al. 1995). This is an important concept because it emphasizes the importance of core or brain temperature in controlling effector output. An increase in hypothalamic temperature produces marked increases in sweating and skin blood flow. Likewise, a decrease in hypothalamic temperature produces shivering and intense cutaneous vasoconstriction. Thus, it seems that it is more specifically the temperature of the brain that is being protected from marked deviations in environmental temperature, rather than the entire core, which includes the abdominal and thoracic cavities. In support of this notion, it is brain temperature that seems to drive sweating, shivering, and skin blood flow responses. Other measures of core body temperature, such as rectal or oral temperature, often are observed to lag well behind a change in sweating or skin blood flow, which suggests that these temperatures are the consequence of a change in effector response rather than the stimulus driving the response.

A crucial point here is that temperature-sensitive and -insensitive neurons are *not* functionally specific; they do not respond *only* to temperature. In some sense, we could call them "polymodal." For example, low glucose excites warm-sensitive neurons and inhibits cold-sensitive neurons, resulting in reduced body temperature. This may help to explain, in part, why the typical alcoholic with a low plasma glucose concentration often is found to be hypothermic. Similarly, hormones can influence the activity of these neurons. For example, estrogen excites warm-sensitive neurons and inhibits cold-sensitive neurons (Boulant and Silva 1989). This may help to explain why postmenopausal women on estrogen therapy have high sweat rates and improved thermal tolerance in the heat.

In humans, the thermal neutral zone is defined as an ambient temperature of 25° to 27°C (77° to 80°F) for a nude individual at rest in the fasted state. Under these conditions, brain temperature is kept constant by delicate changes in cutaneous blood flow without the need to call upon reserve heat-loss or heat-gain mechanisms. Above this ambient temperature range, sweating must be initiated to achieve adequate heat loss to prevent a change in brain temperature, whereas at temperatures below this range, shivering must be initiated to maintain thermal balance.

Surviving the Temperature Straitjacket

Thermal Response to Heat

In a warm environment, the first defense against core temperature rising too far is to increase skin blood flow (provided skin temperature is lower than core temperature), which transfers heat from the core to the skin, where it can be lost primarily by radiation and convection. If core temperature continues to rise, sweating begins and, if evaporated, cools the skin. The primary mechanism that prevents overheating is the evaporation of sweat from the surface of the skin.

Heat, Water, and Shiny Skin

Hair on the general body surface of primitive hominids served to protect them not only from heat loss but also from heat gain. Through evolution, humans lost their general body hair, and with it they lost the ability to protect themselves from the radiant heat of the sun. Parallel to this event, humans developed a unique and remarkable change on the surface of the body not found in any other living mammal: sweat glands. (Some exceptions can be found in primates. For example, the patas monkey has eccrine sweat glands over its general body surface.) Human skin possesses approximately 2 million sweat glands over the general body surface. Activation of these glands expels fluid (sweat) onto the surface of the skin through sweat ducts.

Sweating capacity and active neurogenic vasodilation reach their full potential in humans. Matching these two effector capacities of the thermoregulatory system is important, especially under conditions of great thermal stress like exercise in the heat; otherwise, the huge evaporative capacity (capacity to increase skin blood flow) would be wasted.

Sweat is a liquid composed mainly of water and electrolytes (table 5.1). As this fluid evaporates, it cools the skin. The process of evaporation is analogous to heating water on a kitchen stove. When the water boils, it requires energy to convert liquid water to a vapor. During the evaporation of sweat from the skin surface, the temperature of the skin represents the heat of the kitchen stove and the evaporation of every gram of sweat consumes 0.58 kilocalorie of heat, thereby cooling the skin. When the skin cools, it cools the blood flowing through it; thus heat is transferred from the core of the body to the environment. Heat loss by this mechanism can occur only if there is a vapor pressure gradient from the skin to the ambient air. If relative humidity is 100 percent, no matter how much sweat is produced, none can evaporate and sweat drips from the skin without providing any cooling. In this latter situation, vital body fluids are lost, dehydration occurs, and core body temperature rises at a rapid rate. These conditions have the potential for producing a thermal injury (heat stroke).

Typically, no more than 50 percent of the sweat produced is evaporated, and the maximal amount of heat that can be dissipated by this mechanism is approximately 1,200 kcal/h. The nontrained, nonheat-acclimated individual is capable of sweating at the rate of about 1,500 ml/h. Training and heat acclimatization can increase this figure by about 10 to 15 percent. People like the athlete Derek Clayton, who ran the marathon in 2 hours and 8 minutes, can produce sweat at the rate of 3700 ml/h. This is truly remarkable, for it represents the total volume of plasma in the vascular system. Obviously, the fluid required for sweat production cannot come from the plasma alone. It must be derived from intracellular and interstitial fluid as well.

There are people born without sweat glands (anhidrotic ectodermal dysplasia). They have a difficult time participating in sports that require high rates of heat loss, such as track and cross-country running. A high school miler from Des Moines, Iowa, who was born with anhidrotic ecto-

Table 5.1
Ion concentrations of body fluids

Ion Species	ICF mEq/l	Plasma mEq/l	Sweat mEq/l	Urine mEq/l
Na+	10	140	50	128
Cl-	4	103	50	134
K+	140	4	4	60
Mg++	58	3	2	15

ICF, intracellular fluid.

dermal dysplasia, illustrates an important physiological point. The mile is typically run on a quarter-mile track, and each competitor must complete four laps. This man was able to compete because his teammates formed a bucket brigade and doused him with a bucket of water every quarter-mile of the race, to produce evaporative heat loss. This anecdotal story is extraordinarily relevant because it emphasizes a biological fact: that heat is lost by evaporation and not by the process of sweating itself. In other words, it doesn't make any difference how water reaches the surface of the skin, by sweating or by bucket; it is the process of evaporation that cools the skin, not the process of sweating. An important question is whether these individuals have a high skin blood flow, even though they have no sweat glands.

Heat Acclimatization

Imagine what would happen if a person driving from an air-conditioned office in Los Angeles to Las Vegas in an air-conditioned car suddenly stopped because the engine overheated, and decided to walk in the desert for the next 2 to 3 hours. Without prior exposure to such conditions, this person would experience marked elevations in core and skin temperatures, heart rate would approach maximal values of 200 beats/min, and symptoms of syncope, including dizziness, nausea, and pounding headache, most likely would appear in less than 60 minutes. These physiological responses would occur even if the person had access to sufficient fluids to drink. A heat-acclimatized person ingesting sufficient fluids would be able to walk the 2 to 3 hours in the desert without ill effects. In fact, the rise in core body temperature during such a desert walk would be no higher than during a similar walk in the cool of the evening, when ambient temperature is only 50°F, compared with 120°F during the day (figure 5.4).

How does one become acclimatized to the heat? The process is simple and fast, especially in physically active or endurance-trained individuals. It only requires walking in desertlike conditions 2 to 3 hours per day for 5 to 10 days. At the end of this period, not only does core temperature not rise any higher than it would during the same walk in a cool room, but it is accomplished with a cooler skin, owing to greater evaporative cooling from increased sweating, and a heart rate of only 100 beats/min instead of 200 beats/min (figure 5.4). This is perhaps the most remarkable physiological adjustment that humans are capable of making and is attributed in large part to the evolution of the sweat gland.

As a result of acclimatization, these glands not only are more sensitive to a rise in core temperature, but also secrete a sweat that is almost like distilled water, that is, it has very little salt (NaCl). This is most beneficial because the acclimatized person therefore retains more salt, which stimulates drinking and helps to prevent dehydration. What a marvelous example of physiological brinkmanship! (Additional information regarding the mechanisms of heat acclimatization are presented in chapter 6.)

Thermal Response to Cold

Cold does not exist as a physical entity in and of itself. It is defined as a lack of heat and is not a different form of energy. We are not "pierced" by the cold winds of winter. Cold is simply a sensation that results from heat loss. When humans are exposed to a cold environment, their skin cools and stimulates cutaneous cold receptors, which increase their firing rate as skin temperature decreases. This in turn activates neural pathways to produce superficial vasoconstriction, piloerection or "goose pimples," and possibly shivering. The superficial veins under the skin constrict, and blood returns from the limbs via deep veins, the venae comitantes. This venoconstriction contributes to the lowering of skin temperature, which reduces the temperature difference between the skin and the environment, thereby decreasing heat loss by radiation, conduction, and convection. In very cold weather, vasoconstriction is so effective that tissue insulation can be increased sixfold. This is equivalent to wearing a light wool busi-

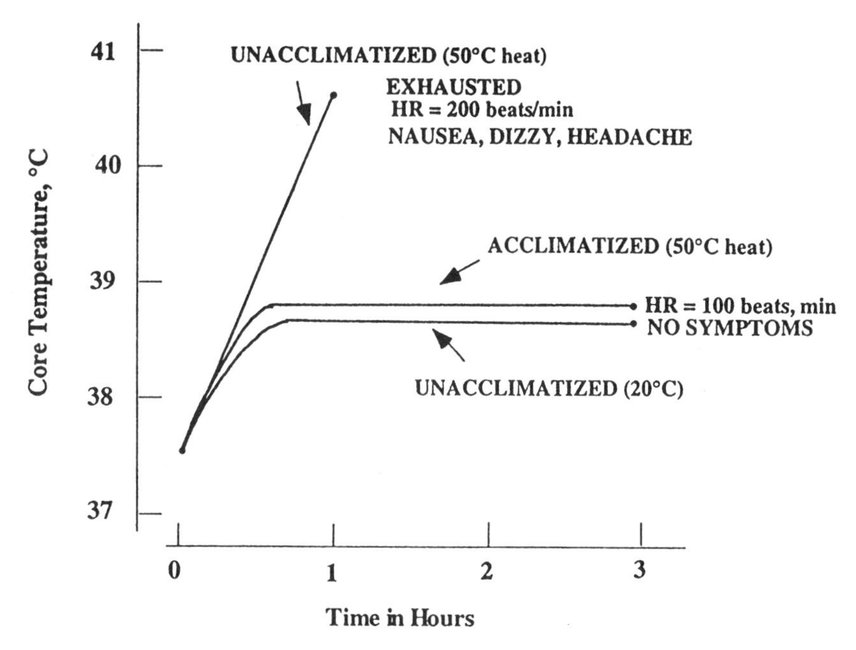

Figure 5.4
Graph of the remarkable physiological adjustments associated with heat acclimatization. Initially, the unacclimatized person can walk only approximately 60 min in desertlike heat (50°C, 120°F) before becoming exhausted. After only 5 to 10 days of repeat exposures, core temperature rises to a plateau no higher than observed during a similar walk in a cool (20°C, 68°F) room.

ness suit. Interestingly, the insulation of the head does not change with environmental temperature because cerebral blood flow is kept constant: at –4°C the amount of heat lost from the head is about half of that produced by the body at rest. *It is wise to wear a hat in the winter.*

The Trembling Body

In addition to reducing heat loss by vasoconstriction, cold can stimulate heat production by activating shivering. Shivering is an involuntary contraction of skeletal muscle that can increase heat production 3- to 4-fold and raise core temperature more than 0.5°C. It consists of alternating contractions of antagonistic muscle groups so that gross body movements do not occur. The efficiency of contraction is minimal, so that virtually all of the chemical energy released in the contractile process is converted into heat. (This inefficient contractile mechanism has been studied in rats. There it accounts for more than 60 percent of the heat production in the cold, which, plus the basal metabolism, may account for more than 75 percent of the total heat produced by these animals in the cold. It also has been suggested that shivering is closely related to the maintenance of muscle tone and may have evolved through the same mechanism.)

Everybody has experienced shivering at least but this response, is not initiated in all people at the same low ambient temperature. In this case obesity has a clear advantage in protecting people from the cold. Obese people start to shiver at a lower skin temperature and maintain a lower blood flow to the skin at any given skin temperature than do lean persons. Based on this concept, is it not paradoxical that women almost always feel colder than men? Because women have a thicker layer of subcutaneous fat than men, one would think that women would endure the cold better than men. So why do they generally feel colder than men?

The explanation is relatively simple. Heat is produced in proportion to body mass. The bigger you are, the more heat you produce. On the other hand, heat loss is in proportion to surface area. The person with the greatest surface area has the greatest heat loss in the cold. We often curl up in the cold to reduce surface area and decrease heat loss. The critical relationship is the ratio of surface area to body mass. Herein lies the answer for women feeling colder than men in cool environments. Women have a much larger surface area/body mass ratio than men. In other words, they have a smaller engine producing heat, but relative to that engine, they have a large surface area to lose the heat they produce. As a result of this heat loss, their body temperature tends to fall, and they feel cold. In addition, women decrease blood flow to the skin more effectively than men do. This reduces their skin temperature more than in men and contributes of the explanation of why they feel colder than men.

Survival in the Cold

Although changes in skin blood flow and surface-to-mass ratio clearly influence responses to the cold, the maximal insulative capacity of the body tissues is not related to these mechanisms, but to body fat content. Let's consider an example. When an ocean liner sinks, who survives? The athlete who has 10 percent body fat and is a competitive swimmer, or the relatively sedentary person who has 20 percent body fat? At first glance, one is tempted to answer that the competitive swimmer survives, because when swimming, he can produce a considerable amount of heat through muscle contraction. However, in the cold, it is not how much heat you can produce, but the amount of subcutaneous fat that you have, that determines survival (figure 5.5). The fall in core body temperature is directly related to mean skinfold thickness. The greater the skinfold thickness, the smaller the decrease in core temperature when the body is submerged in cold water. Therefore, fat is a good insulator and could make the difference between life and death during submersion in cold water. Let's be more explicit: a swimmer with a layer of subcutaneous fat of 3 cm can swim in water at 15°C (59°F) for 6 hours with no decline in core body temperature. On the other hand, the cooling rate of a person with a layer of subcutaneous fat of only 5 mm is very high.

Although some extra subcutaneous fat can make the difference between life and death in cold water, compared with air, fat is a relatively poor insulator. Figure 5.6 shows the insulative value of blubber for an arctic seal compared with the insulation provided by the fur of various animals. Approximately 2.5 inches (6.35 cm) of blubber provides insulation equivalent to about 3 clo (clothing) units. When this clo value is compared with fur, whose insulative value is derived primarily from its ability to trap air, the lemming and squirrel, with only about 0.75 in (1.9 cm) of fur, have the same insulation against cold air as the arctic seal. In cold water, of course, where air cannot be trapped to provide insulation, subcutaneous fat is an important source of insulation. *Thus, in cold air, wearing an appropriate coat with adequate insulation is far superior to building up a layer of subcutaneous fat, as well as being much healthier.*

Human Skin Can "Supercool"!

Should frostbite be a concern as soon as the ambient temperature falls to 0°C or 32°F? At what temperature does human skin freeze, and what factors might influence this freezing temperature? Human skin can "supercool," that is, it can fall well below 32°F before it freezes. In fact, dry skin can supercool to an average of -10.9°C or 11.5°F before freezing. Wetting

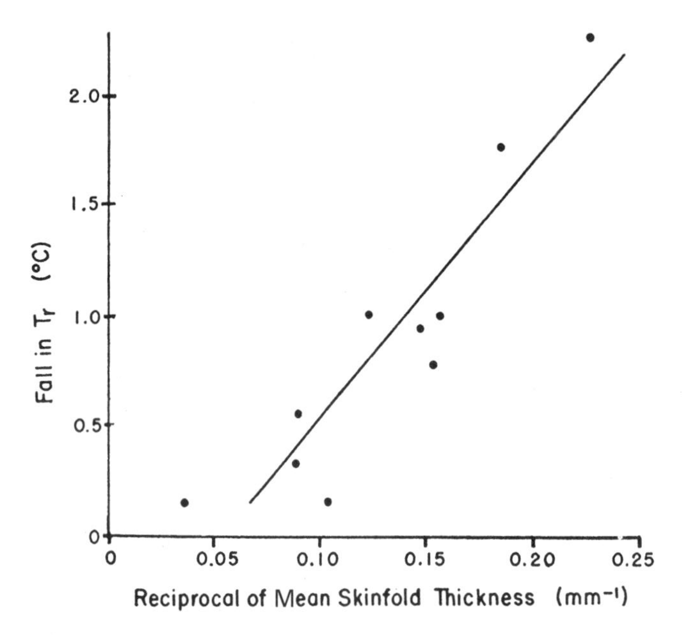

Figure 5.5
Relationship between the fall in rectal temperature and thickness of subcutaneous fat when men were exposed to water at 15°C for 30 min. In cold water, survival depends primarily on subcutaneous fat, which forms a layer of insulation. Heat production is a secondary factor. (Modified from Folk 1974).

the skin causes it to freeze at an average temperature of only -5.4°C or 22.4°F. Water enters the skin through hair follicles, sweat ducts, and intercellular channels. As a result of increasing heat transfer by evaporation, wet skin cools faster than dry skin. However, this faster cooling rate does not explain the higher incidence of freezing. Apparently what happens is that water terminates supercooling by initiating crystallization in the skin, making it more conducive to frostbite (Molnar et al. 1973). *The practical lesson here is to keep your hands out of water if exposure to cold is anticipated.*

Lewis (1941 p. 870) wrote: "It has been found that supercooling displays itself in greater degree in skin that remains unwashed. Washing the skin encourages, while rubbing the skin with spirit and anointing it with oil discourages, freezing. The capacity to supercool greatly would seem to be connected with relative dryness of the horny layers of the skin. It is well known that Arctic explorers leave their skins unwashed."

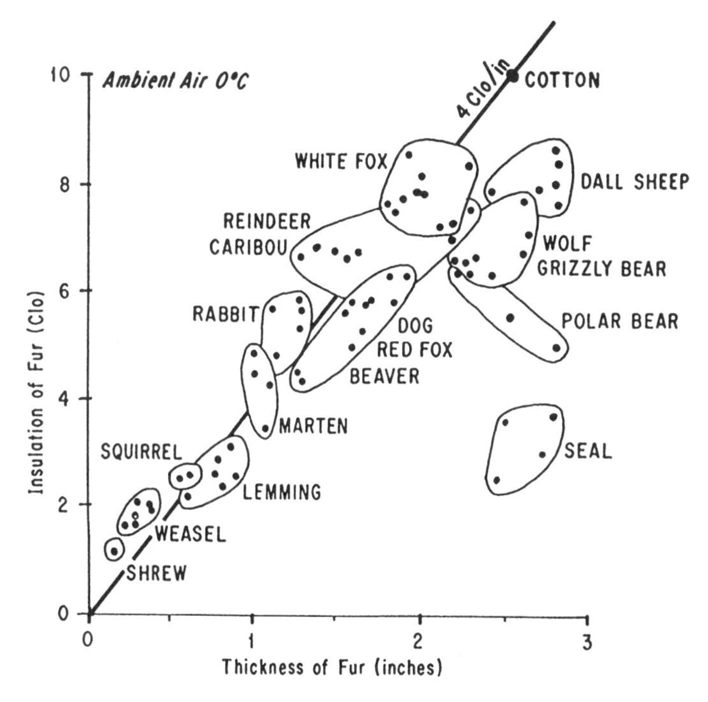

Figure 5.6
Insulation of winter fur. The insulation of fur on different animals is compared with the insulation of 2.5 in of blubber in an arctic seal. (From Scholander et al. In Folk 1974.)

Hats, Clothes, and Boots: Thermal Insulation

The unit of insulation is called the *clo* (short for clothing). One clo of thermal insulation is equivalent to a quarter-inch of clothing or a typical business suit. It's the amount of insulation necessary to keep a resting man, with a heat production of 50 kcal•m^2•h^{-1}, comfortable indefinitely in a 21°C (70°F) environment, with relative humidity less than 50 percent, and wind speed 10 cm•sec^{-1}. The greatest insulator is still air, but when the thickness of a layer of air exceeds 1.27 cm, its insulative value reaches a plateau and may even decline because of the initiation of convective currents. The warmest clothing is that which best traps air, regardless of the material used. Even fine steel wool is as good as cotton wool, as long as the bulk density of the fabric is low. The thermal insulation provided by dead air is 4.7 clo units per inch thickness of dead air, which is very comparable with the best insulating fur of animals (figure 5.6). Thus, the insulative value of fur is proportional to its ability to trap air.

It would be interesting to determine the thickness of insulation or the amount of clothing necessary to withstand different environments and to compare this insulation with the fur and feathers of animals. To determine the number of clo units necessary to withstand a given temperature, the following formula is used:

$$I = \frac{3.1\ (T_s\text{-}T_a),}{\frac{3}{4} M}$$

where I = insulation in clo units; T_s = average temperature of the skin (a comfortable skin temperature is usually taken as 34°C or 93.2°F); T_a = temperature of the air; M = heat production (50 $kcal \bullet m^2 \bullet h^{-1}$ is resting metabolism); 3/4 = 75 percent, or the amount of dry heat loss (25 pecent of our resting heat production is attributed to evaporative cooling).

The amount of insulation required for a resting person to be comfortable in a 20°F environment—6.0 clo units or 1.5 inches of insulation—would be calculated as follows:

$$6.0\ \text{clo} = \frac{3.1\ (93\text{-}20).}{\frac{3}{4}\ (50)}$$

In figure 5.6, this is approximately equal to the insulation provided by the fur of a red fox or a caribou. If this person was walking at a mild pace (5.6 $km \bullet h^{-1}$), heat production would be approximately 150 $kcal \bullet m^2 \bullet h^{-1}$ and the required insulation would be a half-inch (1.27 cm), about 2 clo units. Thus, the extra heat production of mild walking reduced the required insulation for comfort to one-third the value required at rest.

A new index, required clothing insulation (IREQ), has been developed to assess cold stress. It is defined as "the minimal clothing insulation required to maintain body thermal equilibration at a mean skin temperature of 30°C in the absence of any regulatory sweating" (Holmer 1984 p. 1120). It emphasizes the importance of "underdressing" in the cold to avoid problems associated with sweat absorption and condensation. It is easier and safer to add clothing than to remove layers wet with sweat. IREQ defines a steady state condition under which heat loss occurs for a minimal period of time (20–30 min), followed by equilibration of body

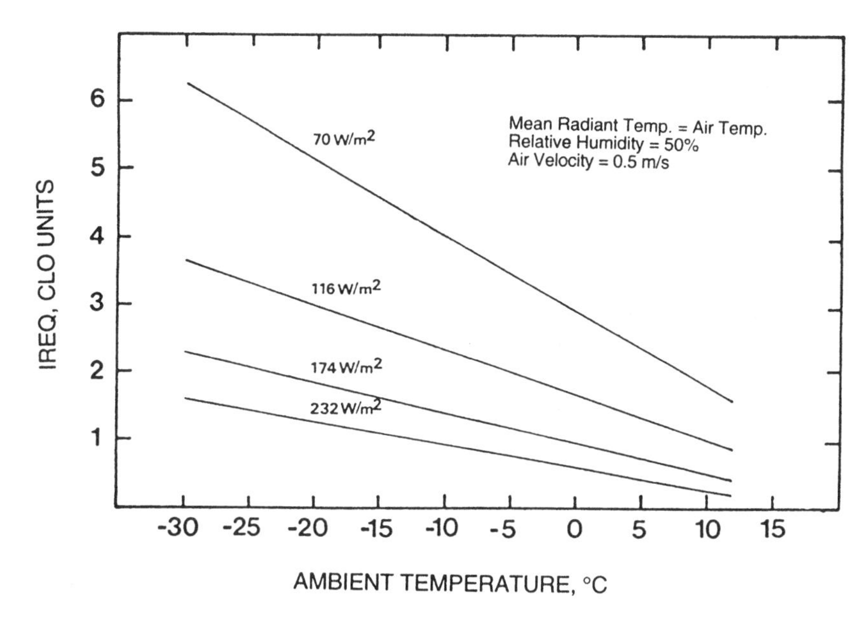

Figure 5.7
Required clothing insulation (IREQ) plotted against ambient temperature at four different levels of metabolic heat production. Clothing insulation decreases as exercise intensity increases from 70 to 232 watts per meter square of surface area (W/m^2.) (From Holmer 1984).

temperature. Figure 5.7 shows that low activity (70 W/m^2) in a cold environment produces considerable cold stress. During exercise, sufficient heat can be produced to balance heat loss. Under these conditions IREQ is small and less dependent on ambient temperature. If the insulation provided by the clothing worn is less than that required to maintain thermal equilibrium, cold exposure must be limited in order to reduce body cooling. This limited duration of exposure for a given set of environmental conditions can be calculated (Holmer 1984).

Cold Acclimatization

Humans can be acclimatized to the cold. In fact, one of the most amazing human capacities is that of sleeping "like a baby," completely nude, at almost freezing ambient temperatures. (This is one of the examples that will be discussed, along with other forms of cold acclimatization, in chapter 6.)

Cold acclimatization is usually associated with an increase in metabolic rate that is not accompanied by shivering. When rats are kept at 5°C, they acclimatize by increasing their metabolic rate 50 to 100 percent, even though shivering decreases. This "nonshivering thermogenesis" may originate in skeletal muscle, but not from muscle contraction, because the increase in heat production with cold exposure still occurs when the muscles are paralyzed with the drug curare. Another fact, contrary to expectation, is that cold-acclimatized people allow their skin temperature to fall to a lower temperature before they begin shivering.

Heat and Cold Acclimatization Together?

An important and intriguing physiological and practical question is "Can heat and cold acclimatization coexist in the same individual, or are they mutually exclusive?" The answer is that they can coexist. The loss of one during a change in season is not due to the presence of the other, but to the absence of an adequate thermal stimulus. Thus, in the transition from summer to winter, the sensitivity of the sweating mechanism decreases, not because the skin becomes cooler during winter months but because the sweat glands are no longer activated by exposure of the skin to a warm environment. Once again, contemplate the effects of prolonged aerobic training. Remember that the most powerful stimulus to activating heat-loss mechanisms is a rise in core body or brain temperature. Thus, the marathon runner who trains in the cold of winter is already acclimatized for work in the heat because he/she elevates core body temperature to high levels with every training session.

Why Not Regulate at 20°C or 40°C Instead of 37°C?

Why Not Regulate at 20°C (68°F)?

Through the process of evolution, all mammals and birds achieved a constant body temperature of approximately 37°C. The small computer in the brain that enabled animals to maintain this constant body temperature was a powerful development of nature because virtually all mammals on Earth, from the arctic fox to the African lion, are equipped with the mechanisms to achieve a fine balance between heat gain and heat loss. Therefore, there should be no question that the regulation of core body temperature at 37°C is the ideal for integrating all physiological mechanisms. If Nature did it, then it should be right. Nonetheless, why is 37°C the ideal temperature, and why is 20°C or 40°C not?

These questions have never received satisfactory answers from biologists. It is surprising how few hypotheses have been proposed to explain this phenomenon. One reason for regulating at 37°C is that it is the optimal temperature for enzymatic function. However, this explanation ignores the fact that more than 50 percent of the body tissues, representing the skin and subcutaneous tissues, are functioning below 37°C. Moreover, the recent discovery of "hyperthermophiles" that can live at temperatures up to 95°C without denaturing would argue against this simple explanation (Gross 1998). Using these observations, Burton and Edholm (1955) concluded that enzymes can adapt to a particular temperature better than the core temperature can adapt to a temperature for optimal enzyme function. He further suggested that homeotherms have selected a high core temperature because it provides maximum stability for temperature regulation.

The argument is as follows. First, a rise in temperature (according to the Law of Arrhenius) increases heat production *and* heat loss. (Temperature influences the rate of chemical reactions, and therefore can influence growth and metabolism. This influence, known as the Law of Arrhenius, is often expressed as the "Q_{10} effect"—the increase in reaction rate for a 10°C rise in temperature. For most living cells, heat production increases 2.3 times with a 10°C rise in temperature. Thus, a 1°C rise in temperature will increase heat production by $(2.3)^{1/10}$ or 8.6 percent. If it is assumed that homeothermy began in a 25°C environment, and that x is body temperature, the difference between core and ambient temperature (the excess temperature) would be (x-25°). A 1°C rise in temperature would then be (x + 1°-25°) or (x-24°). Because heat loss is approximately proportional to the temperature gradient between ambient and core body temperatures, heat loss will increase in the ratio of [(x-24°)/(x-25°)]. Then heat balance should be obtained by equating heat loss and heat production:

$$[(x - 24°)/(x - 25°)] = 1.086\ (2.3^{1/10}).$$

Solving for x yields a temperature of 37°C.

Thus, a temperature that would yield stability between heat production and heat loss would be 37°C. This temperature assumes that heat transfer to the skin from the core is constant and that the heat transfer coefficients for the different avenues of heat loss are constant. However, this approximation is true only for small changes in ambient temperature. Moreover, heat loss is proportional to the thermal gradient from the *skin* to the environment, not from the core to the environment. Burton acknowledges that proof of this theory would not withstand careful scrutiny, but the exercise does suggest that the core temperature of homeotherms may have something to do with the stability of temperature regulation.

Another important consideration is fluid homeostasis. The relationship between animals and their environment is tightly linked to their water balance. Homeotherms evolved in regions of the world where environmental temperature ranged from about 20°C (68°F) to 25°C (77°F). Consider the fluid balance problem that would be created if core body temperature were regulated at 20°C. Under these circumstances, ambient temperature (20° to 25°C) would always be higher than brain temperature. Heat would be gained by radiation, convection, and conduction, and the only avenue of heat loss would be evaporation. The sweating mechanism would be activated continuously and humans would constantly seek water to drink. At a constant brain temperature of 37°C, the core is usually 10°C or more above ambient temperature and the sweating mechanism is not activated.

Why Not Regulate at 40°C (104°F)?

Regulation of brain temperature at this level no doubt would be too close to the denaturation point of protein. (When a protein is denatured, its polypeptide chains uncoil, become tangled with one another, and cannot be separated; hense the denatured protein is insoluble. A good example of protein denaturation is the effect of heat on the protein *ovalbumin* in egg white. When an egg is cooked, ovalbumin, which is normally a soluble protein, becomes denatured and forms an insoluble white coagulum.) Moreover, our mobility and recreation would be markedly limited. Mild activity would produce severe thermal stress, and high intensity exercise, even for short periods of time, would likely result in thermal injury (heat stroke). Sports like tennis, football, basketball, field hockey, and soccer

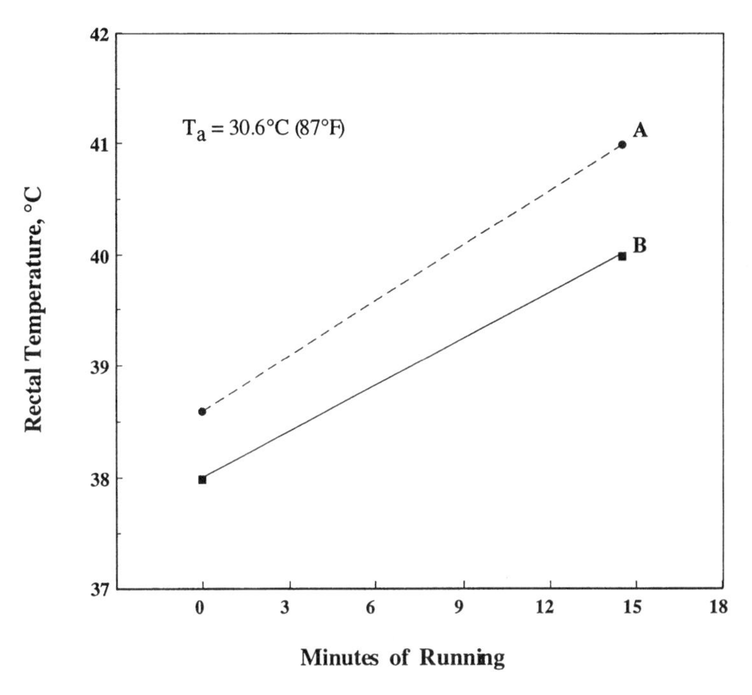

Figure 5.8
Rise in core temperature during exhausting exercise in the heat. This figure illustrates the effect of a competitive 3 mile run on a humid day when the ambient temperature was 87°F. Core temperature was elevated by warm-up before the race. Competitor A achieved a much higher core temperature than competitor B at the end of the race, in part because he had a higher core temperature at the end of his warm-up. (Modified from Robinson 1963.)

would be impossible to play. Even with a core of 37°C, on a warm day with a high relative humidity, core temperature will rise at the rate of 0.3°C per minute during intense exercise (figure 5.8). Thus, within less than 15 minutes, core temperature can rise to over 41°C. As illustrated in figure 5.8, a prolonged or hard warm-up preceding a competitive race will elevate core temperature and contribute to its marked rise during the competition.

Can the regulated temperature be changed? This of course occurs with fever, which is discussed in chapter 8, but aside from fever, is there a temperature other than 37°C that would be better for physiological function? The answer is yes! With acclimatization to heat, core body temperature is regulated at a lower level. Furthermore, after a period of prolonged aerobic exercise, not only is resting core body temperature lower, but both the shivering and the sweating thresholds are reduced. This is associated with improved heat and cold tolerance.

Inferred from these observations is the possibility that primitive man maintained a core body temperature lower than that of the typical sedentary person of today, who is far less tolerant of both hot and cold environments. (Some intriguing insights about primitive man will be discussed in chapter 6).

Why Maintain a Constant Brain Temperature? The Selfish Brain

Neurons are exquisite and delicate cells that require constant and unperturbed surroundings to perform their job. They require chemical constancy and, above all, a warm environment. Neurons are very hardworking cells. They do not seem to complain about their hard and tiring work, but they need help. In fact, for this purpose neurons have their maids, the glia, which are in charge of cleaning up and maintaining their extracellular surroundings.

In the brain of poikilotherms, the neuronal environment was constant with regard to chemicals and hormones, but not temperature. For these animals, neurons in the brain worked hard when temperature was elevated, but slowed their activity when temperature fell. During the transition from reptiles to mammals, these animals had to cope with a new problem, that of producing excess internal heat. Solving this problem required the development of mechanisms for losing heat, and with these mechanisms, a machine, a computer in the brain, to control it. With this computer, and with a constant temperature in the brain, neurons became continually active.

It looks as if in the final stages of mammalian evolution, in the late Cretaceous and Eocene times, the final stage of development of temperature regulation took place. Precisely, it was at the time of growing and further elaboration of the cortical mantle. It has been hypothesized that the evolution of a more complex brain brought a greater need for a constant brain temperature. Recall that the Law of Arrhenius states that reaction rates are a function of temperature. More important, each reaction in the body has an "activation energy" that is temperature-dependent. Activation energy is the energy required to *initiate* a reaction rather than the energy (or heat) produced by a reaction. Moreover, activation energy can vary markedly from one reaction to another.

Imagine the difficulty of performing coordinated movements and complex thought processes if the millions of reactions (with their different activation energies) required to perform these tasks occurred at different times as a function of varying brain temperatures. The emerging concept therefore is that the organizational complexity of the brain and the need for complex interaction of neuronal activity require homeothermy. In support of this concept, it is common knowledge that hypothermia is associated with drowsiness and confusion; and that elevations in brain temperature significantly impair mental performance (Engel et al. 1984; Hancock 1981). Although fever is associated with impaired mental processing, even a nontoxic rise in core temperature produced by exercising in the heat significantly increases the frequency of errors on an attention-stress test. Further evidence supporting the importance of homeothermy for normal brain function is the poor temperature regulation (as an example of neuronal processing) in the newborn compared with an adult (see chapter 4).

6

From Siberia to Africa: Understanding the Extremes

Ducks walk on ice without freezing their feet, rats can live in refrigerators, and some humans can sleep virtually nude at near freezing temperatures without shivering or discomfort. On the other hand, desert ants can maintain body temperatures over 50°C, and elite marathon runners can sustain brain temperatures between 40° and 42°C without heatstroke. How is this possible? One of the most remarkable characteristics of homeotherms is their ability to adapt to changing environments through physiological adjustments and, with more prolonged exposure, anatomical changes. The term used to describe the physiological responses to chronic exposure to environmental stress is *acclimatization*, whereas *acclimation* is used when an organism is exposed to a single stress, usually in a specifically designed environmental chamber where temperature, humidity, oxygen tension, and so on can be controlled and varied independently. During the process of acclimatization or acclimation, if a particular physiological response diminishes compared with the unacclimatized or unacclimated state, the term *habituation* is used. The term *adaptation* usually is reserved for genetic effects produced by natural selection. Often the ability to acclimate to one stress improves one's ability to cope with other stresses. This is called *crossacclimation.* A good example of crossacclimation is the effect of endurance training on heat and cold acclimation.

Cold Acclimatization

Many researchers believe that humans are not capable of acclimatizing to the cold because humans do not expose sufficient body surface to the cold for a sufficient period of time to produce an adequate stimulus to elicit an adaptive response. With this notion in mind, (a) are there any human populations that expose their whole body to the cold sufficiently to produce acclimatization, and (b) is it possible to produce local acclimatization to the cold as a result of limited exposure of body regions? For example, if your hands are usually exposed to the cold, do they exhibit greater cold tolerance than the hands of someone who usually wears gloves?

There are only three possible physiological adjustments that can be made to the cold: (a) decreased heat loss, achieved by reducing the surface area of skin exposed to the cold, growing hair for insulation, and/or decreasing the temperature gradient from the skin to the environment by reducing skin temperature; (b) increasing heat production by shivering or nonshivering thermogenesis; or (c) allowing body heat to decrease to the level of hypothermia. This last change is not synonymous with hibernation. Shivering is a specialized form of neuromuscular activity, whereas nonshivering thermogenesis is a generalized increase in metabolism of several or all bodily tissues. Is there any evidence for whole-body cold acclimatization in humans?

The Eskimo: A Frustrated Expectation

The concept that humans are incapable of acclimatizing to the cold is supported by the majority of the data collected on the Alaskan Eskimo. Contrary to popular opinion, the physiological responses of Eskimos to environmental stress are similar to those of Europeans. Both groups shiver and sweat at the same skin temperatures. From a physiological viewpoint, these responses should not come as a great surprise, since both groups maintain virtually the same skin temperature despite the fact that the Eskimo lives in the Arctic. Eskimos simply wear appropriate clothing to protect their skin from the cold. On the other hand, Eskimos generate 30 to 40 percent more heat from nonshivering thermogenesis than do Europeans not typically exposed to the cold. This elevation in metabolism is attributed to the high-protein diet consumed by the Eskimo and not to cold exposure. When the diet of the Eskimo is limited in protein content to that of a white Caucasian, the elevation in metabolism is reduced to that of the European.

Although Eskimos do not expose their whole body to the cold, they do expose their extremities, and show evidence of local acclimatization. For example, Eskimos have a much greater capacity to keep their hands warm than do Europeans. This is attributed to hand blood flow, which was found to be twice that of medical students living in a temperate climate. When their hands were immersed in 5–10°C water, Eskimos maintained higher hand temperatures and showed greater increases in blood flow, especially to the fingers, than the Europeans. Even Eskimo children, with a smaller hand volume and greater surface-to-mass ratio than their elders, had warmer hands than adult white Caucasians.

This local acclimatization generally is found in fisherman who repeatedly have their hands in cold water, and is attributed to a blunted vasoconstrictor response in the skin areas exposed to cold. What is intriguing is that the blunted vasoconstrictor response is observed not only in the exposed limb but also in the contralateral nonexposed hand, indicating that this response is mediated by the brain (Eagan 1963).

The *Ama*: Diving in the Cold for a Living

In contrast to the Eskimo, the women of Korea and Japan who dive to harvest plant and animal life year round in waters that reach 10°C during the winter provide the best evidence of cold acclimatization in humans. When physiologist Suki Hong studied these women in the early 1960s, their basal metabolic rate during the winter months, when they were diving in very cold water, was significantly elevated above values observed during warmer months; this provided the crucial evidence for cold acclimatization. However, years later, when these women were restudied, they no longer showed this enhanced metabolic capacity. This deficit could not be attributed to diet, age, or the loss of physical conditioning. What happened?

Traditional Divers

These Korean and Japanese women divers are called *ama*. They begin diving at the age of 11 or 12, and may continue to the age of 65 (Hong and Rahn 1967). They dive both winter and summer and, if pregnant, are known to work up to the day of delivery. In the late 1950s, when Hong

Figure 6.1
Korean diving woman wearing a traditional bathing suit made of light cotton. (From Hong and Rahn 1967.)

began his classical studies, these women wore only cotton bathing suits (figure 6.1), even during the winter, when the air temperature was near freezing and seawater temperature was only 10°C. In the winter, a diver worked 1 or 2 15- to 20-min shifts; even this short time was sufficient to lower rectal temperature from 37°C to 34.8°C (Hong et al. 1987). This level of hypothermia also has been observed in Australian Aborigines (see below) and Channel swimmers. A typical dive lasted about 60 sec and consisted of a 30-sec dive followed by a 30-sec rest interval in the water. During the summer, when air temperature is 25–30°C and water temperature is 22–27°C, divers worked 2 1-hour shifts and 1 2-hour shift. Each diving sequence was terminated when oral temperature reached about 34°C. The lowest oral temperature recorded was 32.5°C (Hong et al. 1987).

In addition to monitoring changes in rectal temperature, cold stress was estimated from increments in oxygen uptake (Hong et al. 1987). During a winter diving sequence, heat loss was estimated to be 600 kcal, compared with 400 kcal, during a summer diving shift. Considering that a diver may work one or two shifts in the winter and three shifts in the summer, the average heat loss was estimated to be 1,000 kcal per day during each season. To compensate for this caloric deficit, divers consumed approximately 3,000 kcal in food each day, compared with only about 2,000 for nondivers.

To oppose heat loss and the consequent lowering of body temperature, the *ama* increased their basal metabolic rate by 30 percent during the winter (figure 6.2), whereas nondivers in the same community exhibited no seasonal changes. This remarkable adjustment is attributed primarily to whole-body exposure to cold. As noted above, the Alaskan Eskimo has a metabolic rate comparable with that of the *ama* during winter, but the Eskimo's high metabolic rate is attributed to a very high-protein diet. By comparison, the protein content of the *ama* diet is low, and cannot account for their elevated basal metabolic rate.

In addition to elevating heat production, another important adaptive mechanism of the *ama* against the cold is increasing body insulation. A major factor in body insulation is the layer of fat under the skin; because women have more subcutaneous fat than men, they excel as divers. However, the *ama* as a group are relatively lean individuals and the

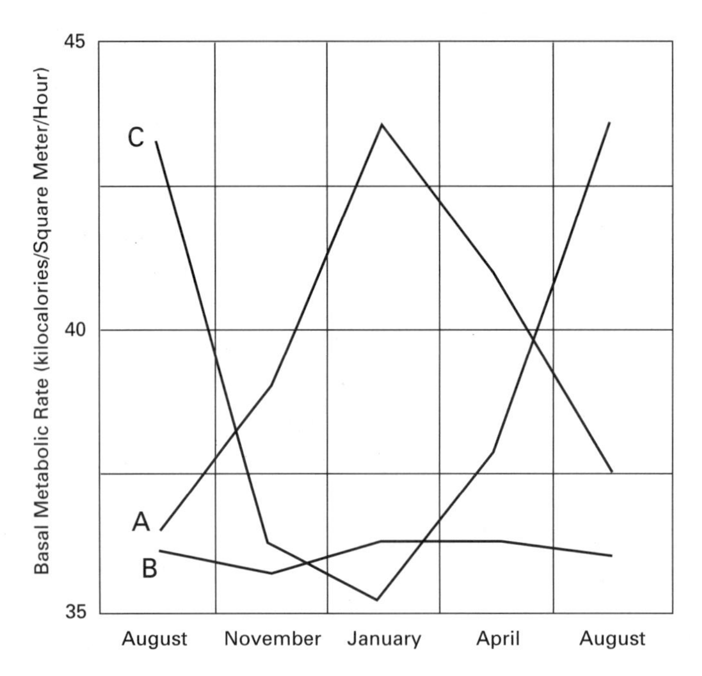

Figure 6.2
Basal metabolic rate of *ama* women (A) increases in winter and decreases in summer. In nondiving women (B), basal metabolic rate is constant throughout the year. Curve (C) shows the mean seawater temperature in the diving area of Pusan harbor for the period covered by the other measurements. (Modified from Hong and Rahn 1967.)

amount of subcutaneous fat they possess typically is lower than that of nondivers, yet they lose less heat under conditions of cold exposure. How is this possible? The answer relates to the control of peripheral blood flow, especially to the limbs, and the measurement of *functional* insulation. This measure is obtained by submerging subjects in cold water for several hours and recording their core temperature and heat production. Using these results, maximal body insulation is calculated from the equation

Maximal body insulation = [T_C-critical water temperature]/Rate of skin heat loss,

where T_C is core temperature (rectal temperature), critical water temperature is the lowest water temperature the subject can tolerate for 3 hours

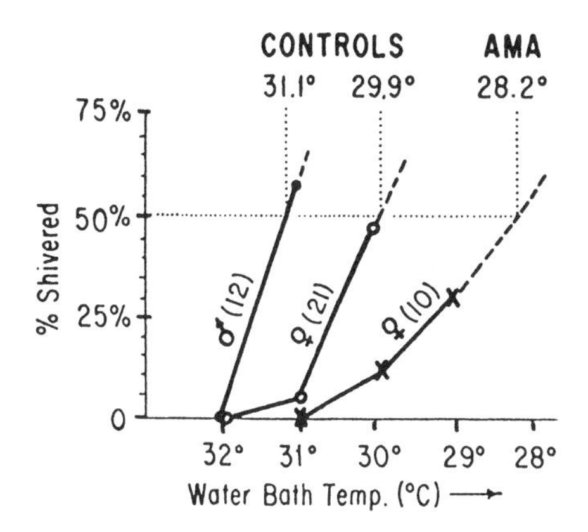

Figure 6.3
Shivering threshold of Korean diving women is shifted from 32°C to 31°C. The difference between control groups (men, nondiving women) and the *ama* is more apparent at 50 percent shivering. The *ama* become cold-acclimatized in winter and shiver at a lower temperature than nondivers and men. (From Hong 1973.)

without shivering, and the rate of skin heat loss is equal to metabolic rate (corrected for respiratory heat loss) plus net loss of stored heat during submersion. The outcome of these experiments indicated that the *ama* (a) experience less heat loss than nondivers with the same thickness of subcutaneous fat; (b) lose about half of their subcutaneous fat in the winter, presumably because their caloric intake does not keep pace with their heat loss; and (c) shiver at a significantly lower water temperature than nondivers. The water temperature at which 50 percent of the *ama* shivered was 28.2°C, compared with 29.9°C for nondivers, and 31.1°C for men (figure 6.3). This greater insulative shell of the *ama* is attributed to the suppression of shivering, but, it could be a vascular response. The latter could be due to an enhanced vascular constriction of the limbs and/or a more effective countercurrent heat exchange in the limbs. Because there are no significant differences in heat flux from the limbs or in limb blood flow when comparing the *ama* with nondivers, it has been concluded that the greater insulation is due to a more efficient countercurrent heat exchange system (Hong et al.1969).

In addition to the response to whole-body submersion, another test of cold tolerance is to submerge the hand in 6°C water and to monitor the local vascular response. Contrary to the attenuation of finger vasoconstriction observed in arctic fishermen and Eskimos, the *ama* showed greater vasoconstriction than nondivers. This response is attributed to the whole-body cold exposure of the *ama* (versus with only local cold exposure experienced by Eskimos and arctic fishermen). The latter groups expose their hands to the cold, but the rest of their skin is maintained at a relatively high temperature and their core temperature changes little.

Contemporary Divers

In 1977, the *ama* began wearing wet suits to combat the cold stress. This change provided the rare opportunity to study the time course of the decay of acclimatization. Diving in wet suits certainly made the work of the divers more comfortable and productive. Rectal temperature fell only 0.6°C during a diving sequence in the winter, compared with a fall of 2.2°C when the *ama* wore a cotton bathing suit. In a wet suit, mean skin temperature was as much as 10°C higher, and heat loss was reduced to only 37 percent of what it was when the *ama* wore a cotton bathing suit. As a result of the reduced cold stress, there was no seasonal change in metabolic rate. Basal metabolism was no different in the *ama* compared with nondivers. When measures of critical water temperature were made for a given thickness of subcutaneous fat, divers showed a marked change in 1983, compared with values observed in the 1960s. The difference in critical water temperature between the *ama* and nondivers in the 1960s was as much as 4°C (Hong et al. 1987), compared with only 1°C in 1981 and no difference in 1983 (table 6.1). When subjected to the cold-hand test (described above), the more intense vasoconstrictor response of traditional divers was lost. This supports the notion that the latter response of was associated with whole-body cold exposure rather than cooling of the hands alone. Thus, the *ama* who lost their cold acclimatization blunted the stimulus to acclimatization by wearing artificial insulation instead of preserving their own insulation.

This return of thermoregulatory function to control levels after several years of diving in a protective garment provides evidence that the cold exposure experienced by traditional divers indeed produced cold acclimatization.

Table 6.1
Thermoregulatory functions in traditional and contemporary divers

Year	Basal Metabolic Rate (%)	Maximal Body Insulation (%)	Q_{finger} (%)	Critical Water Temp (°C)
		Traditional divers		
1960–1977	+30 (in winter)	+20	–30	–4
		Contemporary divers		
1980	0	0	–20	–2
1981	0	0	0	–1
1982	0	0	0	–0.2
1983	0	0	0	0

Note: Basal metabolic rate, maximal body insulation, and Q_{finger} (finger blood flow during immersion of hand in 6°C water) are expressed as percent deviations from respective control values (+ and - denote higher and lower than control, respectively). Critical water temperature indicates absolute value relative to control (- denotes lower than control values).
From Hong et al. 1987.

The Australian Aborigine: Sleeping Nude in the Cold

There are a number of ways to test human tolerance for cold exposure. Perhaps the most widely used is to submerge the subject in cold water or expose the subject to cold air, and to determine the threshold skin temperature for the onset of shivering. A less popular and more time-consuming test is the "cold bed test." Basically, the individual is confined to bed in the cold and subjected to a variety of physiological and behavioral tests; the most important one is whether the person is able to sleep. When white Caucasians are tested, they feel cold, shiver, toss and turn, cannot get comfortable, and cannot sleep. The Australian Aborigine, on the other hand, sleeps like a baby. Most interestingly, when noncold-acclimatized white Caucasians participate in long-term aerobic training, they display many of the same physiological adjustments as the Australian Aborigine. What are these physiological adjustments, and how are they developed in endurance athletes?

In contrast with the *ama*, the Aborigines show little or no increase in metabolism during cold exposure. Thus, their ability to tolerate the cold

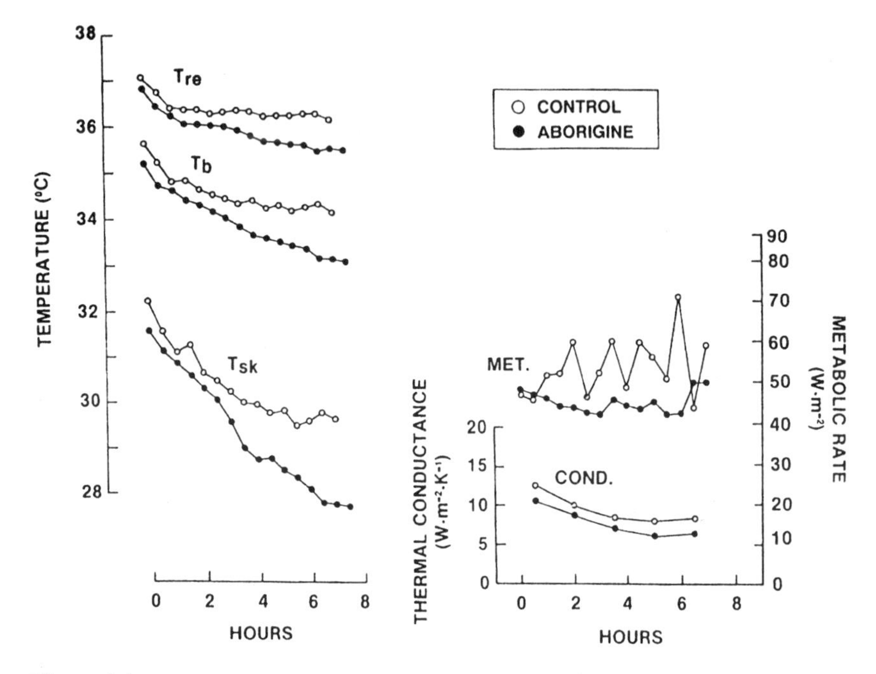

Figure 6.4
Average thermal and metabolic responses of seven Australian Aborigines and six control white Caucasians during a night of moderate cold exposure (ambient temperature about 5°C). These tests were performed in a portable field environmental chamber to obtain a suitable and reproducible sleeping environment. (Redrawn from Hammel 1964.)

is not due to a metabolic adaptation. When the responses of a group of control white Caucasians were compared with those of a group of central Australian Aborigines during the night, rectal, skin, and mean body temperatures, metabolism, and skin conductance were all lower in the Aborigines (figure 6.4).

Because of their higher metabolic rate, the heat content of the control group changed very little, but the heat content of the Aborigines fell continuously throughout the night. The rate at which heat content fell diminished as the gradient between skin temperature and air temperature decreased. The conductance of heat from the body core to the skin was about 30 percent less in the Aborigines than in the controls, even though mean skinfold thickness (3.04 mm) and percent body fat (17.3 percent) were considerably lower in the Aborigines than in the whites.

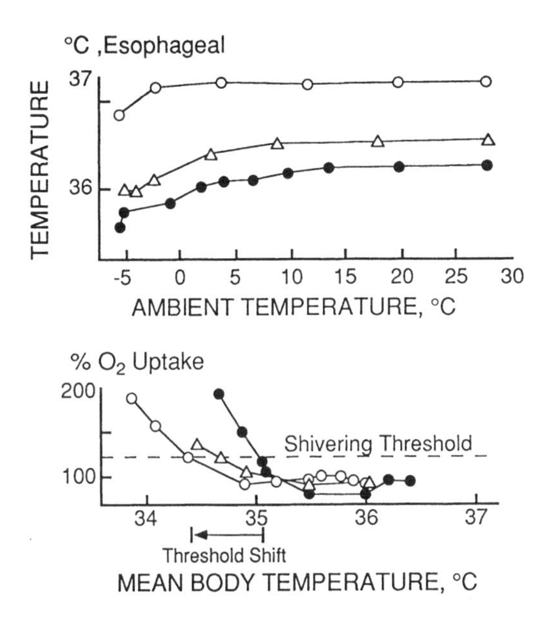

Figure 6.5
Comparison of esophageal temperature and oxygen uptake responses of one subject to cold exposure before (•) and after 1 year (O) and 1.5 years (Δ) of marathon training. Upper panel shows the change in esophageal temperature as ambient temperature was reduced from 29°C to –5°C. Lower panel shows the corresponding change in oxygen uptake as a function of mean weighted body temperature. Although it appears that the threshold for an increase in O_2 uptake did not change, note the decrease in shivering threshold and diminished rise in oxygen uptake with endurance training. (Redrawn from Baum et al. 1976.)

Corresponding values for the controls were 5.36 mm and 22.3 percent, respectively (Hammel 1964). This form of cold acclimatization, whereby (a) heat content of the body decreases, allowing core temperature to fall, and (b) conductance of heat to the skin is diminished, thereby increasing the thickness of the body shell, has been referred to as "insulative hypothermia." Considering the difficulty of obtaining food in their environment, this form of acclimatization adopted by the Australian Aborigine is certainly more energy-efficient compared with metabolic acclimatization adopted by Korean and Japanese diving women. The mechanisms responsible for adoption of these different strategies for coping with the cold by the *ama* and the Aborigines are not understood.

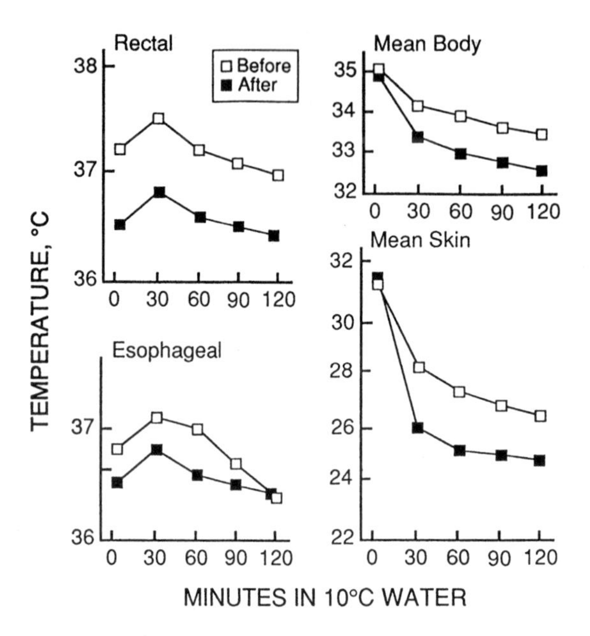

Figure 6.6
Rectal, esophageal, mean body, and mean skin temperatures of a lean subject exposed to 10°C air for 2 h, before and after training. (Redrawn from Kollias et al. 1972.)

What is intriguing is that endurance training, which improves heat tolerance, also improves cold tolerance. It is understandable that exercise training, even in a cool environment—which markedly elevates core body temperatures, stimulates the sweating response, and increases blood flow to the skin—would improve heat tolerance. But why does endurance training improve cold tolerance? Figure 6.5 illustrates some of the physiological responses of a subject exposed to the cold before and after 1 and 1.5 years of marathon training (Baum et al. 1976). The two responses after training were similar. Esophageal temperature was considerably lower after training as ambient temperature was reduced from 29°C to –5°C, and the cold-induced rise in heat production (O_2 uptake) occurred at a considerably lower mean weighted body temperature. In fact, O_2 uptake did not reach 200 percent until mean body temperature had fallen almost 1°C more after training than before training. Mean body temperature at rest and at the onset of shivering and sweating were all progressively reduced as the level of fitness (based on marathon time) improved. Because cold sensation and thermal discomfort occurred at a lower mean body temperature, these observations indicate that exercise training reduced the thermoregulatory set-point.

In a related study, 9 weeks of training reduced core, skin, and mean body temperatures, metabolic rate, and tissue heat conductance during exposure to a 10°C environment for 2 hours, compared with values observed before training (figure 6.6). In addition, physical conditioning produced an average body weight and fat loss of 1.8 and 2.9 kg, respectively. Skinfold measurements indicated a loss of subcutaneous fat. Because subcutaneous temperature rose moderately after training, whereas skin temperature fell, the greater temperature gradient across the skin indicated that greater peripheral vasoconstriction may result from physical training. This implies that physical training increases insulation of the skin.

Collectively, these findings mimic those observed in the Australian Aborigine and suggest that exercise training produces a form of cold acclimatization described by Hammel (1964) as "insulative hypothermia." Thus, endurance training results in *crossacclimatization*. The latter term is used because training not only reduces the thresholds for shivering and sweating, which might be characterized by the term *habituation*, but also increases maximal sweating and the upper limit to which core temperature can be driven without thermal injury.

What is the rationale for this type of physiological adjustment to training? Recall that the hypothalamus is a complex structure responsible for integrating homeostatic functions including energy, heat, and water balance. Endurance training will impact on all of these functions. The decrease in shivering threshold may be a manifestation of fuel conservation. In addition, because resting body temperature also is reduced, training may help to lengthen the time interval before body temperature rises to a level that would impair performance. Consider the following calculation. An endurance athlete weighing 70 kg can maintain a heat production of 700 watts. After one hour of maintaining thermal balance, it has been shown that heat dissipation can fall about 10 percent as a result of a decrease in skin blood flow shortly before exhaustion (Adams et al. 1975). Heat storage under these conditions would be 70 watts • 3600 seconds/70 kg = 3.6 kJ • kg^{-1}. Because the specific heat of the body tissues is 3.48 kJ • kg^{-1} • °C, mean body temperature would rise about 1°C in 1 hour. Thus, based on this calculation, a decrease in resting body temperature of only 0.1°C would prolong performance time by about 6 minutes.

The brain's regulation of adaptive responses to the cold is poorly understood and can vary with different neurotransmitters, brain sites, and animal size, age, development, and species. Based on data from guinea pigs of the same strain, age, and sex, the body temperature threshold for the onset of shivering depends on a balance between norepinephrine (NE) and serotonin (5-HT) inputs impinging on the anterior hypothalamus from nuclei in the lower brain stem (Zeisberger and Roth 1996). In fact, cold-acclimated animals have a shivering threshold that is approximately 1°C lower than warm-acclimated animals. It was concluded that warm-acclimated animals have greater activity in norepinephrine pathways, because when NE was injected into the anterior hypothalamus, the shivering threshold moved to a higher core temperature in cold-acclimated, but not in warm-acclimated, animals. On the other hand, blockade of NE receptors reduced the shivering threshold much more in warm-acclimated than in cold-acclimated animals.

Heat Acclimatization

In chapter 5, we described what would happen if an unacclimatized person was stranded in the desert during the summer and had to walk for several hours to reach civilization. Humans are unique because, unlike other vertebrates, they have the tremendous capacity to dissipate heat by secreting sweat over the general body surface. However, even if this person had adequate water to drink, the consequences of such a trek could be fatal. The cardiovascular and thermoregulatory systems in an unacclimatized person are unable to respond to such stress. Trying to supply oxygen to the legs to continue walking and blood to the skin to dissipate heat places a heavy burden on the cardiovascular system, especially when blood volume is decreasing because of excessive sweat production. As a result, heart rate rises to 200 beats/min and core temperature to 40°C or higher, sweating may decrease as a result of sweat gland fatigue, and it is common to experience nausea, dizziness, headache, blurred vision, and loss of coordination. What is even more remarkable is how quickly these symptoms and signs of severe stress are abated with repeated short-term

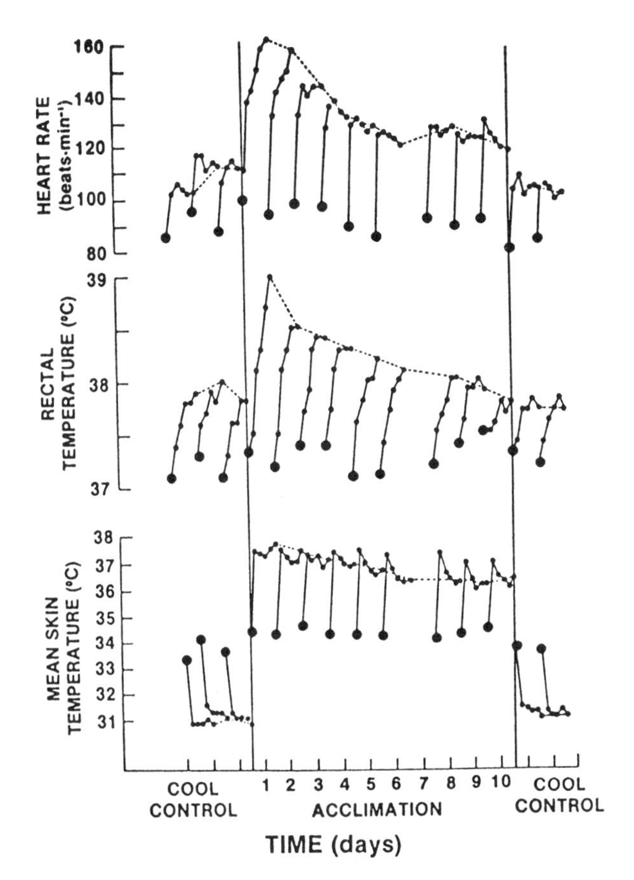

Figure 6.7
Changes in heart rate and rectal and mean skin temperatures in a cool room before and after a 10-day regime of heat acclimation to a hot, dry environment (50°C). Each day, activity consisted of 5 10-min periods of treadmill exercise alternated with 2-min rest periods. The initial value each day is indicated by a large dot and the end of each exercise period by a small dot. The final values of each day are connected by a dotted line. (Taken from Eichna et al. 1950.)

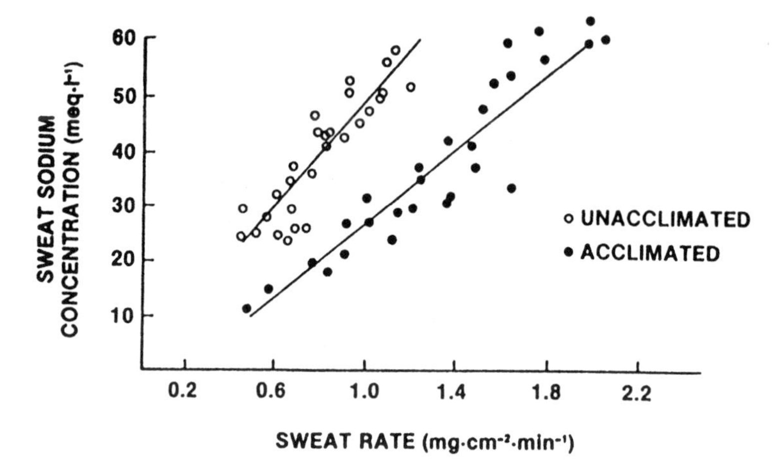

Figure 6.8
Effect of heat acclimation on sweat sodium concentration as a function of sweat rate. (Modified from Alan and Wilson 1971.)

(60–90 min) exposures (figure 6.7). After only 10 days of intermittent walking in the heat for 50 min, the rise in rectal temperature in 50°C (120°F) heat was no greater than in a cool 25°C (77°F) environment. Acclimatization to heat is one of the most remarkable physiological adjustment that humans are capable of making. How does this happen? Is the brain the orchestrator?

In humans, almost 80 percent of heat acclimatization occurs within 3–4 days. In this brief period, there are significant increases in body fluid compartments, with the secretion of aldosterone playing a crucial role in retaining sodium from the kidneys and the sweat glands causing a dilute sweat (figure 6.8), which helps to maintain extracellular fluid volume and the drive to drink. If the effects of aldosterone are blocked by the drug spironolactone, the deleterious symptoms and signs outlined above reappear. On the other hand, if the person is maintained in salt balance, the role of aldosterone in the acclimatization process is diminished.

As the acclimatization process continues beyond day 4, blood volume and extracellular fluid volume tend to decrease, but remain elevated above values observed on day 1. In the meantime, other remarkable changes are taking place. Not only does core temperature decrease, but heart rate and skin temperatures also decline. The decline in skin temperature is the result of an increase in evaporative cooling. It increases the core-to-skin temperature gradient, which facilitates heat transfer to the skin and heat loss, so that core temperature declines. The decrease in core temperature reduces skin blood flow, which leads to an increase in the heart's stroke volume. Heart rate decreases because stroke volume is elevated and core temperature is lower. Cardiac output remains unchanged. The primary stimulus for this cascade of events is probably the rise in core body temperature. This concept is supported by the increase in heat tolerance produced by exercising in a cool environment and elevating core temperature (Gisolfi 1973).

Figure 6.9 shows the changes in forearm blood flow, which is used as a measure of skin blood flow, and sweating in a group of subjects who were exposed to exercise in the heat before undergoing a 2-week physical training program in a cool environment (pre-exercise), after 2 weeks of training (post-exercise), and after 5 days of repeated heat exposure (post-heat) to bring about heat acclimation. Two weeks of physical training produced little change in forearm blood flow (left panel), but the sweating response at any given esophageal temperature was greater, that is, the sensitivity of the response increased. With heat acclimation, the sensitivity of the blood flow response increased, as indicated by an increase in the slope of the relationship. The effect of acclimation on the sweating response was more dramatic. The sensitivity (slope) of the response did not change compared with the response to physical training, but the threshold esophageal temperature for the onset of sweating was significantly reduced from 37.5°C to 37.0°C. Although 2 weeks of training produced only a change in the sensitivity of the sweating response, more prolonged training, as practiced by endurance athletes, would likely produce the same response as heat acclimatization. This is because endurance training, even in a cool environment, produces a high and sustained elevation in core temperature.

Skin biopsies obtained from monkeys before and after heat acclimation showed that after acclimation, sweat glands increased in size, produced more sweat, and became more efficient in that they produced more sweat per unit length of secretory coil (Sato et al. 1990). When similar biopsy

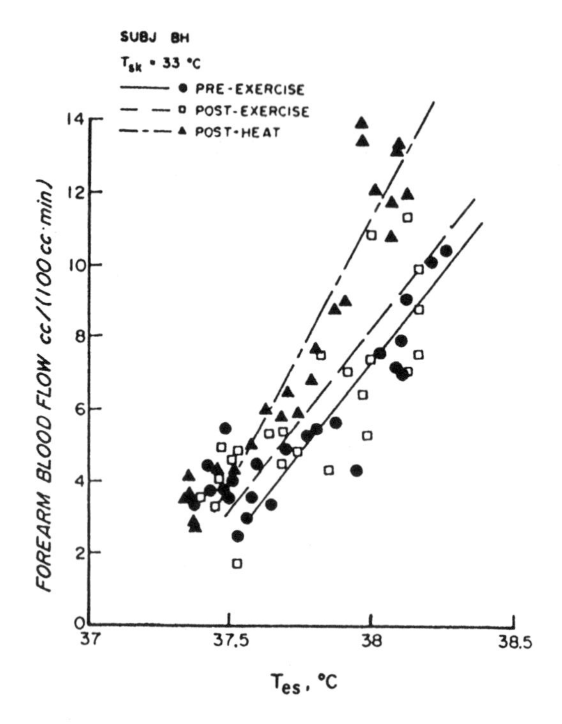

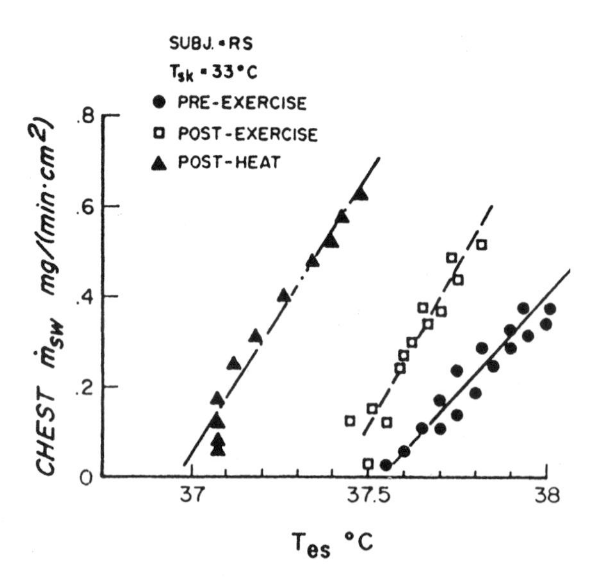

Figure 6.9
Effects of two weeks of training and heat acclimation on forearm blood flow and sweating on the chest (as a function of esophageal temperature, T_{es}). (Taken from Roberts et al. 1977.)

specimens were obtained from human subjects who characterized themselves as poor or heavy sweaters, the latter group showed significantly larger glands that produced more sweat and were more sensitive to methacholine, which simulated their response to the in vivo neurotransmitter acetylcholine (figure 6.10).

The decrease in the threshold for the onset of sweating and the increase in skin blood flow are considered to result from changes in norepinephrine- and dopamine-mediated heat loss. Evidence to support this concept is presented in figure 6.11, which shows that heat acclimation increases norepinephrine-induced peripheral heat-dissipating capacity. This increase is shown by the significant change in slope of the dose-response curve. The marked shift to the left of this curve suggests that heat acclimation produces a change in receptor sensitivity (Christman and Gisolfi 1985). Interestingly, Pierau et al. (1994) found a significant decrease in warm-sensitive neurons in the preoptic anterior hypothalamus (POAH) of heat-acclimated animals and no cold receptors. This would imply a change in synaptic input impinging on warm- and cold-sensitive neurons to modify their firing rates.

How does the process of heat acclimation affect the synthesis of heat shock proteins (HSP), and what is the effect of heat acclimation on the synthesis of HSP in the brain, particularly at sites known to subserve temperature regulation? The role of HSP in cell survival and cell homeostasis will be introduced in chapter 7. In that discussion, it is noted that animals residing in warm environments have high levels of HSP. This higher level of HSP in proportion to the temperature of the environmental niche is interpreted as a mechanism to protect the organism from thermal stress. In support of this concept, preliminary data from heart tissue indicate that warm-acclimated rats have significantly higher basal levels of HSP than animals living under cool ambient conditions. Whether similar elevations in HSP concentrations occur in other organs and in the brain, especially areas participating in temperature regulation, is not known.

There are conflicting data on the thermal sensitivity of brain tissue. Brinnel et al. (1987) reported that neurological disorders and brain lesions can occur at relatively low levels of hyperthermia. Moreover, the LD_{50} (lethal dose 50), the core temperature (and presumably the brain temperature) at which 50 percent of individuals will suffer a heatstroke, is

A.

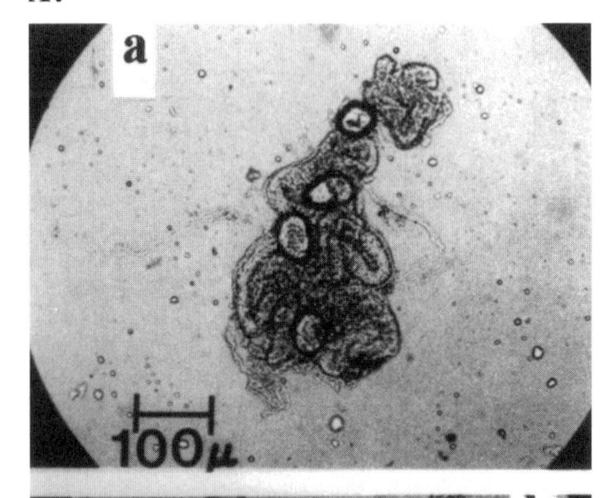

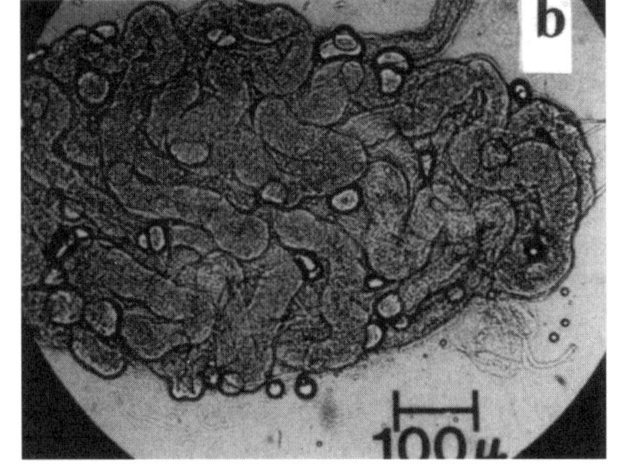

B.

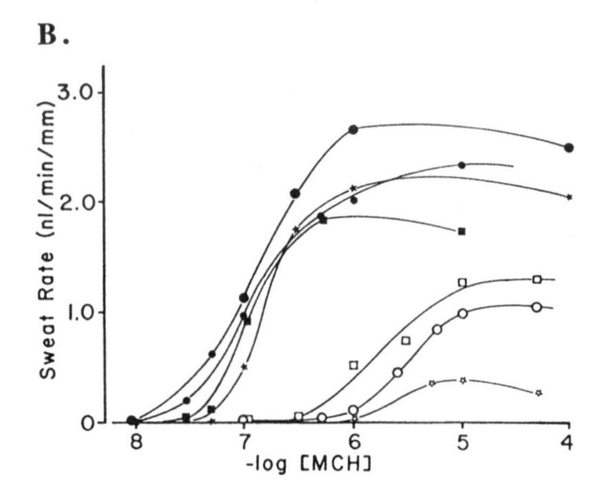

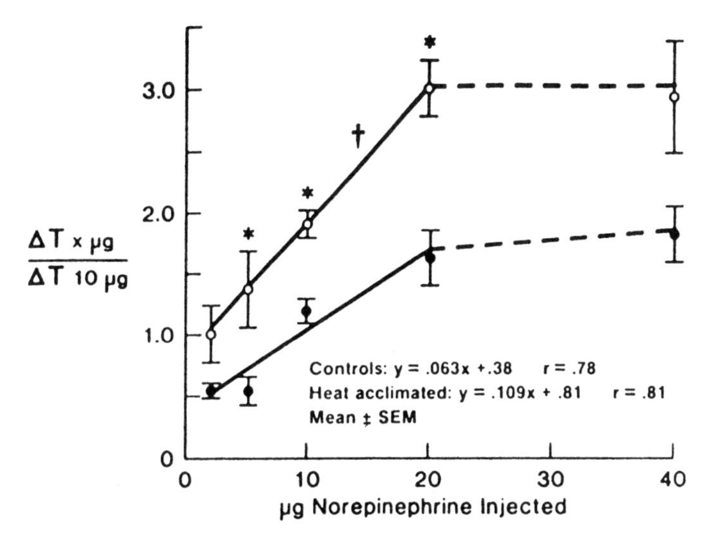

Figure 6.11
Dose-response curve for norepinephrine microinjected into the preoptic anterior hypothalamus before (closed circles) and after (open circles) heat acclimation in the rat. The change in colonic temperature with each dose of norepinephrine has been normalized by dividing the change in temperature with each dose by the temperature change produced by the 10 µg dose (ΔT x µg/ΔT 10 µg). *Significantly greater ($P<0.05$) than control response at that dose. †Linear portion of dose-response curves significantly different ($P<0.001$ split-plot analysis of variance). (Modified from Christman and Gisolfi 1985.)

Figure 6.10
(A) Sweat glands from biopsy specimens of poor sweaters (open symbols) (a) and heavy sweater (closed symbols) (b). It is presumed that the differences in sweat gland size and sensitivity are associated with the level of fitness of these subjects. (B) Dose-response curves to methacholine (MCH) of 4 subjects who were heavy sweaters (closed circles) and 3 subjects who were poor sweaters (open circles). (Modified from Sato and Sato 1983.)

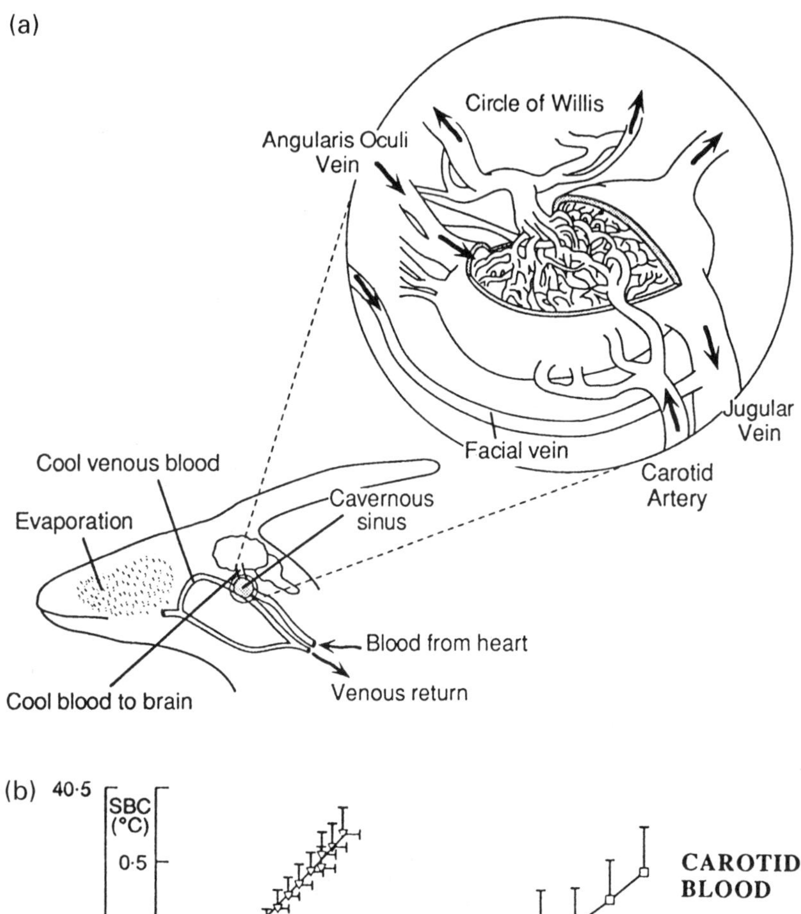

(a)
Circle of Willis
Angularis Oculi Vein
Jugular Vein
Facial vein
Carotid Artery
Cool venous blood
Cavernous sinus
Evaporation
Blood from heart
Venous return
Cool blood to brain

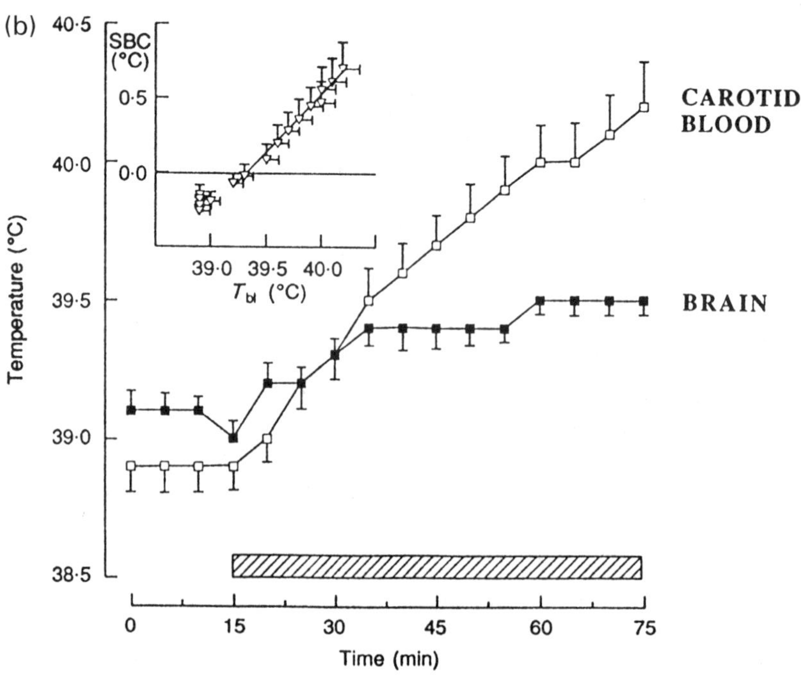

(b)
SBC (°C)
0·5
0·0
39·0
39·5
40·0
T_{bl} (°C)
Temperature (°C)
40·5
40·0
39·5
39·0
38·5
CAROTID BLOOD
BRAIN
0
15
30
45
60
75
Time (min)

only 40.4°C (Hubbard et al. 1976). On the other hand, two species of antelope, Grant's gazelle and the oryx, survive desert exposures by allowing their trunk temperature to rise to over 46°C. Is this also the temperature of the brain? If not, is it possible for some animals, and even humans, to cool the brain?

Rats, Cats, Dogs, and Humans: Who Can Cool the Brain?

It's a hot summer day on the Mojave Desert. A jackrabbit, with an implanted radio telemeter to measure body temperature, is observed to have a resting core temperature of 41°C. A dog spots the rabbit and immediately gives chase. In 10 min, the body temperature of the jackrabbit, which is sprinting to avoid the dog, rises to 43°C. Unable to hide, it continues to run and elevates its body temperature to 44°C, then suddenly dies. Why didn't the dog also become hyperthermic and die? The dog, being much larger, you reason, must have produced more heat and probably had a higher body temperature than the rabbit. The answer, at least in part, is that the dog was able to cool its brain. How is this possible?

In 1966, a group of investigators discovered that when goats became hyperthermic, their brain temperature rose less than the rest of their body (Taylor 1966). They termed this phenomenon "selective brain cooling" (SBC). Subsequent studies revealed that SBC occurs in artiodactyls (even-toed mammals with hooves: pig, hippopotamus [4 toes], antelopes, camels, deer, giraffe, cow [2 toes], dogs, cats, and other species. The mechanism that allows SBC to occur is attributed primarily to the presence of a *carotid rete* (figure 6.12), a network of medium-sized arteries embedded in the cavernous sinus at the base of the brain. The cavernous

Figure 6.12
(A) Schematic diagram of countercurrent exchange in the brain of an antelope. Venous blood draining the nasal and buccal cavities flows back to the heart through the cavernous sinus, cooling arterial blood destined for the brain. The magnification shows the carotid rete as a network of small arteries formed from the carotid artery within the cavernous sinus. (B) Carotid blood (T_{bl}) and brain temperature of a goat during exercise, illustrating selective brain cooling (SBC). Inset shows SBC (the difference between blood and brain temperatures). The cross-hatched bar represents 60 min of level treadmill exercise at 4.8 $km \bullet h^{-1}$. (Modified from Taylor and Lyman 1972; Baker 1993; Baker and Nijland 1993.)

sinus serves as a countercurrent heat exchanger because warm carotid blood destined for the brain passes through the cavernous sinus, which receives cooled venous blood returning from the mouth and nasal passages en route to the heart.

When dogs are running at high speeds, their brain temperature can be 1.5°C below their body temperature (figure 6.13). The mechanisms responsible for this high rate of brain cooling during exercise include (a) panting; (b) increased secretion of nasal and salivary glands, which accelerates the rate of evaporative cooling; and (c) an increase in nasal and oral mucosal blood flow. This magnitude of brain cooling also has been observed in Thomson's gazelle (an African antelope weighing 7–12 kg that has a carotid rete). A 1°C difference between brain and carotid blood temperatures has been observed in the horse (McConaghy et al. 1995). Animals with a high exercise tolerance seem to have a greater capacity for SBC. Neither the horse nor the rabbit has a carotid rete, but the horse has a much greater exercise capacity than the rabbit and shows greater SBC during exercise than at rest. The rabbit, with a much lower exercise capacity, shows similar levels of SBC during rest and exercise. The capacity to increase blood flow to the upper respiratory passages, as well as to respiratory and skeletal muscles, during exercise may be as important as having a carotid rete (McConaghy et al. 1995). These marked differences between brain and trunk (blood) temperature have led to the interpretation that SBC protects the brain from thermal damage. Because the brain is considered to be more vulnerable to heat than other organs (Burger and Fuhrman 1964; Brinnel et al. 1987), SBC is considered to have survival value under conditions of hyperthermia.

However, when brain and body temperatures were recorded by radiotelemetry in free-ranging wild animals in their natural environment, SBC was observed under normothermic conditions but was not observed when the animals were chased by a high-speed helicopter so that scientists could retrieve their data recorders (figure 6.14). These observations were made in the springbok *Antidorcas marsupialis* (a small South African antelope weighing 25–30 kg) and the black wildebeest (a medium-sized South African antelope weighing 130 kg). The observation that SBC did not occur under conditions of flight to escape predators, and the finding that some animals can sustain a high brain temperature without injury,

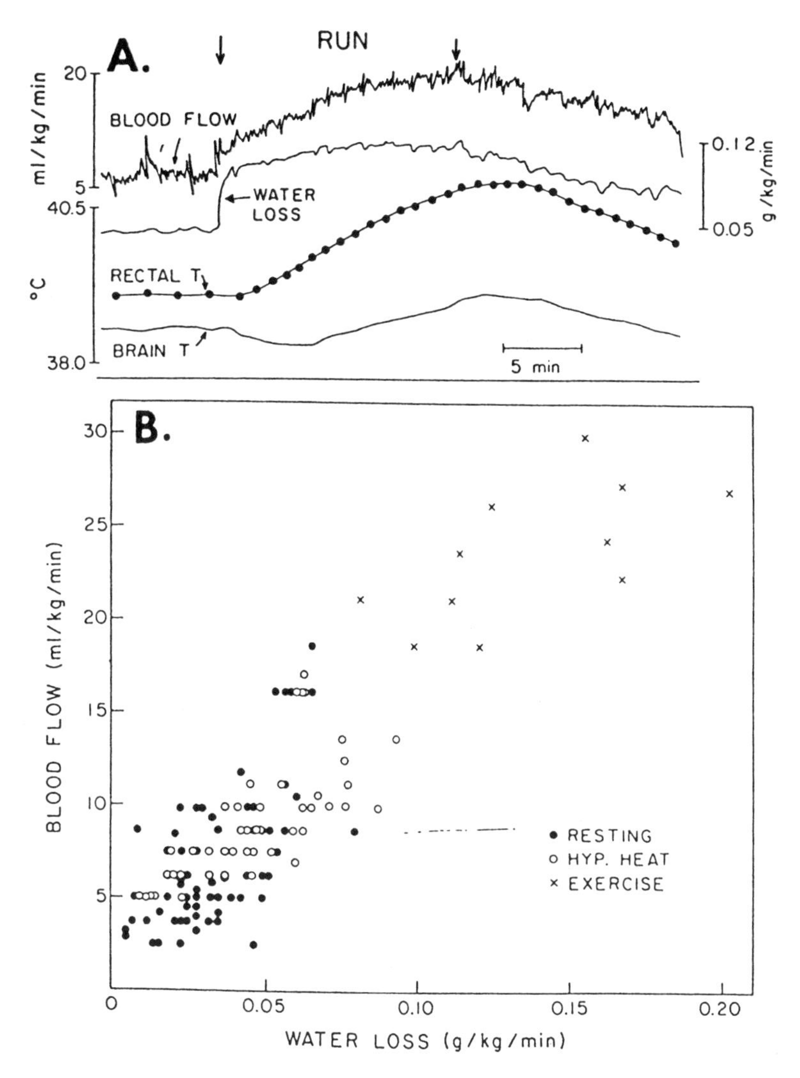

Figure 6.13
Relationship between cephalic blood flow and upper respiratory evaporation in the dog. (A) Carotid blood flow (implanted ultrasonic probe), evaporative water loss (flow-through mask), rectal temperature, and brain (hypothalamic) temperature during exercise. In the period marked RUN (between the vertical arrows), the dog ran at 7.5 km/hr on a 20 percent slope. Ambient temperature was 25°C. (B) Steady-state levels of carotid blood flow and evaporative water loss in a dog at rest at ambient temepratures from 25°C to 45°C (black circles), during heating of the hypothalamic thermosensitive zone (white circles), and during the last minute of 15 min of exercise at different workloads and different ambient temperatures. Carotid blood flow and evaporative water loss were highest during heavy exercise in a warm environment. (Taken from Baker 1982.)

A

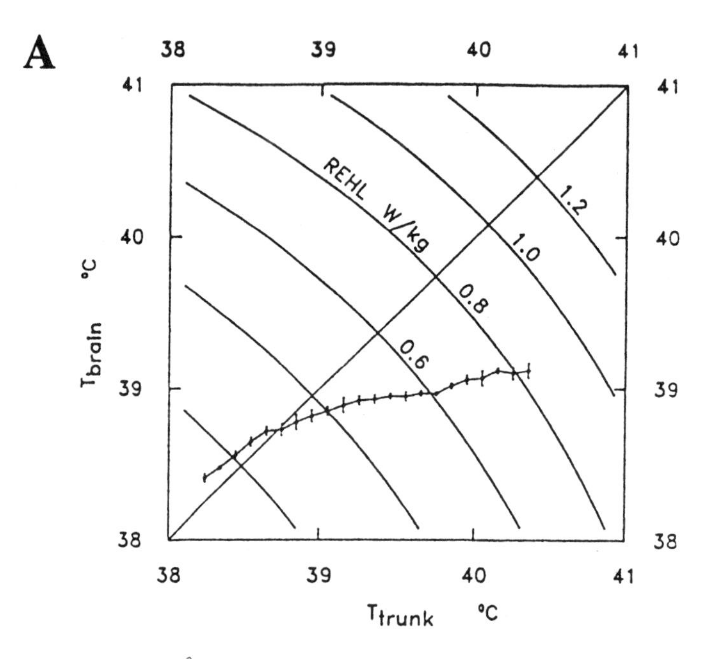
REHL W/kg
1.2
1.0
0.8
0.6
T_{brain} °C
T_{trunk} °C

B

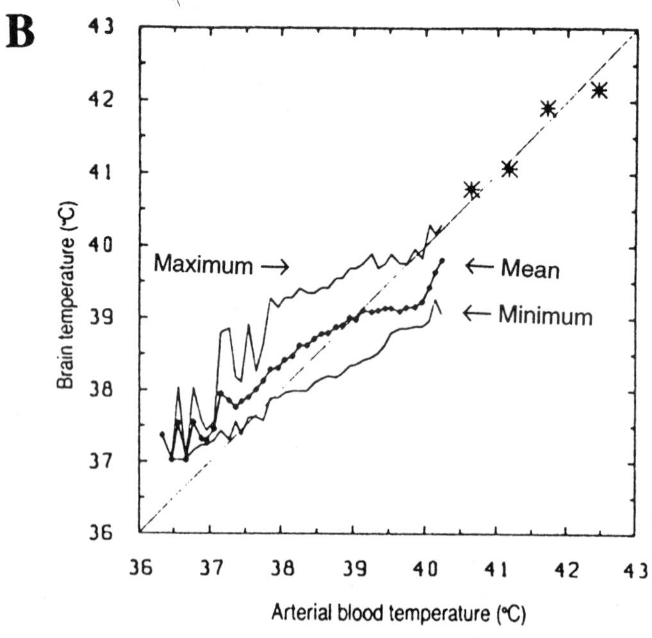
Maximum →
← Mean
← Minimum
Brain temperature (°C)
Arterial blood temperature (°C)

has led to an alternative interpretation of SBC: that it serves to modulate thermoregulation rather than to protect the brain against heat injury. This interpretation is supported by the finding that SBC is observed under hyperthermic conditions as well as under normothermic circumstances (figure 6.14). Thus, SBC may be considered a thermoregulatory effector mechanism similar to shivering and sweating, with the reservation that its thermoregulatory effect is limited to the brain. Measurements made in free-roaming antelopes showed that SBC occurred during the day under conditions of intense radiant heat. Respiratory evaporative heat loss (REHL) was reduced because the brain temperature was lower than the trunk temperature. This strategy conserves body fluids. In dehydrated goats, SBC is maintained at reduced levels of REHL (Baker and Nijland 1993).

What does the hypothalamic thermostat sense in animals with a carotid rete? If a large number of temperature sensors are in the brain, why would nature evolve a system that cools this area of the brain, thus preventing it from detecting upward deviations in core body temperature? As stated in chapter 3, thermosensors are located in the spinal cord and throughout the body core. Afferents from these trunk sensors contribute equally to the control of respiratory heat loss if brain and trunk temperatures are identical (Jessen and Feistkorn 1984). When brain and trunk temperatures are independently manipulated, brain temperature provides exclusive control over SBC (Kuhnen and Jessen 1991). The combination of a high trunk temperature and a low brain temperature results in panting but not SBC. On the other hand, the combination of a high brain temperature and a low trunk temperature produces SBC but not panting.

Figure 6.14
Schematic diagram illustrating when SBC is turned "on" or "off" in an animal under free-ranging conditions. (A) At rest or during moderate activity under conditions of low environmental heat stress, SBC is evoked (middle), which inhibits brain temperature sensors from detecting the rise in body temperature. Consequently, respiratory evaporative heat loss (REHL) is driven primarily by trunk temperature sensors and the rise in REHL with increasing body temperature is attenuated (bottom). (B) During high sympathetic activity—the animal running for its life—SBC is suppressed, that is, brain temperature equals trunk temperature (middle), and REHL rises with a steeper slope because it is now driven by both trunk and brain temperature sensors. (From Jessen 1998.)

Thus, in the hyperthermic animal, the brain can offset (separate) its own temperature from the rest of the body. SBC can reduce the drive on thermoregulatory effectors that are activated by input from the core and brain. However, when SBC is active and brain temperature is lower than trunk temperature, REHL is reduced and the trunk contributes more than the brain to the drive for heat loss.

As illustrated in figure 6.14, SBC in the resting animal reduces REHL, thereby conserving body fluids, whereas SBC is not operative when the animal is running for its life and a maximal drive for REHL is desirable. An important feature of the system, as described by Jessen (1998), is that SBC is not mandatory; that is, cool venous blood from the nasobuccal cavities can return to the heart via the angularis oculi vein through the cavernous sinus and lead to SBC, or it can return through the facial vein directly to the jugular vein and prevent SBC (figure 6.12) (Johnsen et al. 1987). These veins contain sphincters that are richly endowed with sympathetic fibers. Thus, it may be possible for the sympathetic nervous system to activate or suppress SBC by manipulating blood flow via these two routes because the angularis oculi sphincter has α-adrenergic receptors that are normally relaxed, whereas the facial vein sphincter has β-adrenergic receptors that are normally constricted. Thus, low sympathetic activity could promote cool venous flow through the cavernous sinus, leading to SBC, whereas high sympathetic activity could constrict the angularis oculi sphincter, dilate the facial sphincter, and direct cool venous blood away from the cavernous sinus, thereby preventing SBC.

What can we conclude from these findings? How can we reconcile the observations in the wild with those in the laboratory? Jessen and colleagues (Jessen 1998) have elegantly shown in field studies that antelope appear to turn off SBC under "fight or flight" conditions. However, the animals in this situation were being chased by helicopter, which no doubt produced maximal emotional stress in addition to their running at maximal speeds. SBC under these conditions could have been overwhelmed by intense heat production and thus not have been apparent. When exercise studies using the same species were performed under laboratory conditions, the intensity of the exercise was not maximum and the animals were not subjected to the emotional stress associated with being chased in the wild. The level of sympathetic outflow had to be markedly greater in the wild than in the laboratory, and this could help to explain why SBC occurred in the laboratory but not in the wild. The finding that SBC occurs under normothermic conditions in an arid environment points to the thermoregulatory benefits of this phenomenon. By limiting the panting response and allowing trunk temperature to rise, the animal conserves vital body fluids in an environment where water is limited. The storage of heat that results from radiant heat gain and elevates trunk temperature during the day is dissipated in the cool air of the evening hours. Thus, SBC economizes the use of panting by reducing the respiratory rate for a given change in brain temperature. However, under the most severe conditions—running from a predator—high sympathetic stimulation precludes SBC, hypothalamic temperature rises with trunk temperature, and panting is driven to its highest level (figure 6.13B).

Is SBC active in fever? If SBC functions to protect the brain under hyperthermic conditions, one might expect it to be fully activated during fever. The results in this area are controversial and seem to vary with species, which may indicate that SBC depends on the nature of the carotid rete. When goats were studied during fever and the results were compared with data obtained in a nonfebrile state, it was concluded that fever inhibits SBC by elevating its threshold and reducing its slope. These observations support a thermoregulatory role for SBC.

Does SBC occur in humans? Humans do not have a carotid rete and do not pant, but some investigators believe that significant cooling of the brain can occur nevertheless (Cabanac 1986). This possibility is a controversial issue that depends on whether or not brain temperature can be estimated accurately (Nadel 1987; Wenger 1987; Brengelmann 1993; Cabanac 1993). The temperature that has been used to estimate brain temperature has been tympanic membrane temperature (T_{tym}). The question is "Does T_{tym} track esophageal temperature (T_{es}), which is usually acknowledged as the best measure of core body temperature, or does T_{tym} fall below T_{es} under conditions of heat stress and during manipulations such as face-fanning, and in so doing provide evidence of SBC?" Before answering, it is critical that T_{tym} be measured accurately. If it is not, spurious data will be collected and incorrect interpretations will be made. If the temperature-sensing device employed to measure T_{tym} is not

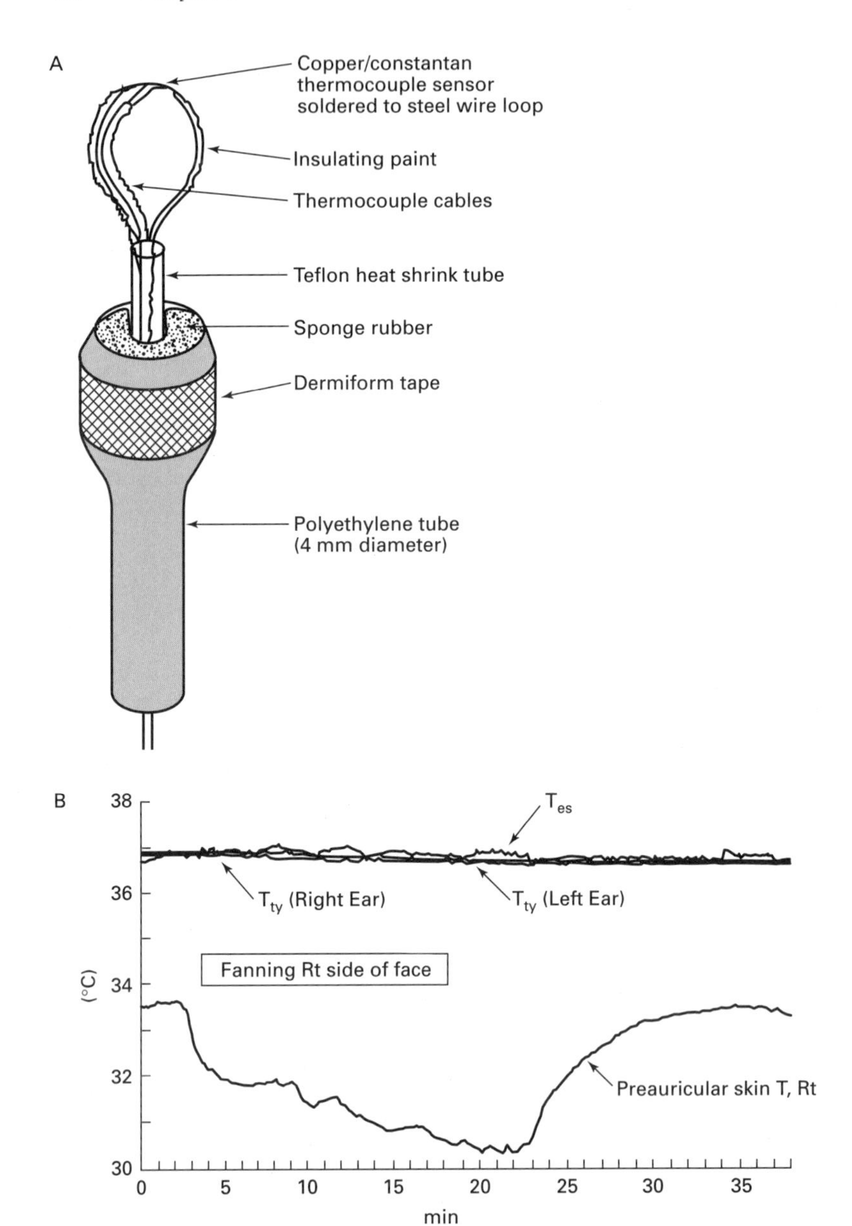

Figure 6.15
(A) Tympanic membrane thermocouple. (B) Effect of fanning the right side of the face on tympanic membrane temperature (T_{ty}) in both ears and on esophageal temperature (T_{es}). (Modified from Sato 1996.)

properly insulated and placed on the membrane, it will sense a temperature that is lower than T_{es}. Figure 6.15 shows that when a spring-loaded thermocouple is used (to ensure that its tip is indeed on the tympanic membrane), T_{tym} temperature will closely track T_{es} changes. Moreover, during fanning and heating of the face, skin temperature can be changed markedly without producing an alteration in T_{tym}. These observations and the finding that the interpeak latencies of acoustically evoked brain stem potentials (whose changes reflect changes in brain stem temperature) (Jessen and Kuhnen 1992) do not change with face-fanning argue against significant SBC in humans. It has been calculated that blood flow through the cavernous sinus must increase four- to fivefold (Nielsen 1988) or that venous blood in the sinus must be 119°C cooler than the arterial blood (Wenger 1987)for SBC to occur in humans. However, proponents of SBC in humans argue that heat exchange within the cavernous sinus represents only one mechanism for lowering brain temperature. Direct conductive cooling also could occur from the outer layers of the brain, where a 0.4°C temperature gradient exists from 4 cm to 2 cm deep (Whitby and Dunkin 1972). The lower temperature in the outer layers presumably is the result of heat loss from the surface of the head.

Nielsen (1988) found that T_{tym} while subjects were bicycling outdoors (wind velocity 5–7 m•sec^{-1}) was about 1.0°C lower than T_{eso}. This difference was not observed when subjects were cycling indoors under still-air conditions at the same ambient temperature. This temperature difference between indoor and outdoor exercise could be attributed to (a) conductive cooling of the eardrum, (b) constriction of arteries supplying the tympanum, or (c) local countercurrent heat exchange (figure 6.16). Conductive cooling was considered unlikely because the ears were plugged with wax and covered with cotton and earmuffs. Pulmonary ventilation and environmental conditions were similar indoors and outdoors, thus eliminating a difference in evaporative heat loss as a possible explanation. Because face temperature was 8°C lower outdoors than indoors (increasing the convective heat transfer coefficient 15–20 times over still-air conditions), countercurrent heat exchange could occur between the external jugular vein and external carotid artery, internal jugular vein and common carotid artery, or the cavernous sinus and internal carotid

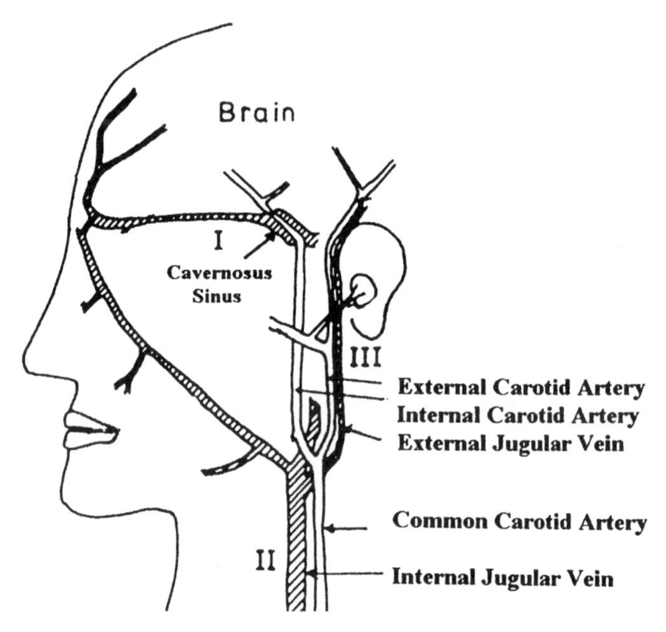

Figure 6.16
Potential sites for countercurrent cooling of arterial blood supplying the hypothalamus and tympanic membrane. Site I: cavernous sinus–internal carotid artery. Site II: internal jugular vein–common carotid artery. Site III: external jugular vein–external carotid artery. (From Nielsen 1988.)

artery. Compressing the internal jugular vein had no effect on T_{tym}, so heat exchange between the internal jugular vein and common carotid artery was eliminated as an explanation.

Using anatomical dimensions, blood flows, physical properties of blood and tissues, and conventional heat exchange equations, Nielsen (1988) calculated that cooling of arterial blood flowing through the cavernous sinus could account for only 0.2°C, far less than the 1°C difference observed between T_{eso} and T_{tym}. This led to the following conclusions: (a) the lower T_{tym} observed outdoors resulted from local cooling of regions supplied by the external carotid artery (which supplies 80 percent of the blood flow to the tympanic membrane); (b) T_{tym} is not a good measure of hypothalamic temperature (see also Brengelmann 1987); and (c) the T_{eso}–T_{tym} difference is not a measure of brain cooling.

Thus, the debate continues. If SBC occurs in humans, it probably is limited. Moreover, recent data indicate that heatstroke patients can sustain temperatures in excess of 41°C and that heat acclimation and endurance training not only enhance heat-dissipating capacity but also increase the core and, presumably, the brain temperature that can be tolerated without thermal injury (see Hales et al. 1998 for review).

7

The Burning Brain

Each year approximately 2 million people from over 80 nations gather to make the great pilgrimage to Mecca, the hajj. This is not simply a visit to a holy place. Every Muslim, everywhere in the world, unless physically or economically unable to do so, is obligated to perform the hajj at least once in a lifetime. It is the fifth of the five fundamental "Pillars of Islam" (Long 1979). In the Quran it is written that the first house of worship founded for mankind, the Standing Place of Abraham, is in Mecca (Long 1979). The timing of the hajj is absolutely fixed on the eighth, ninth, and tenth days of the last month of the Muslim calendar. Because the hajj occurs according to the Muslim lunar year, which is 11 days shorter than the solar year, at the end of about a 34-year period it has fallen in each of the 12 solar months. Thus, it falls in the summer (April–September) and winter in 17-year cycles. Aside from cholera, heat is the major source of curable medical problems. This is because of the desert climate found throughout the Hejaz. In the summer, maximum temperature in the shade has been recorded at 52.2°C (126°F) (Long 1979). Even in the winter, when the climate is balmy, constant exposure to the sun can be debilitating, especially for the aged and young. In the June 1959 hajj, 454 people died of heat illness (Long 1979). Heatstroke is a completely preventable condition, but the circumstances surrounding the hajj continue to make it a major challenge for those responsible for administering this annual event. This type of heatstroke is known medically as "classical heat stroke" (table 7.1).

Let's consider a very different scenario that can lead to greater morbidity and mortality than classical heatstroke. Imagine trying to run 135 miles beginning at 6:30 in the morning, when ambient temperature is

Table 7.1
Characteristics of classical and exertion-induced heatstroke

Characteristics	Classical	Exertional
Age	Older	Young
Occurrence	Epidemic form	Isolated cases
Pyrexia	Very high	High
Predisposing illness	Frequent	Rare
Sweating	Often absent	May be present
Acid–base disturbance	Resp. Alkalosis	Lactic Acidosis
Rhabdomyolisis	Rare	common
DIC	Rare	Common
Acute renal failure	Rare	common
Hyperuricemia	Mild	Marked
Enzyme elevation	Mild	Marked

DIC, disseminated intravascular coagulation.
From Knochel 1989.

37.8°C (100°F). By midafternoon, ambient temperature rises to 55°C (131°F), and asphalt (ground-level) temperature is 82–93°C (180–200°F). These were the conditions under which competitors participated in the 1996 Badwater, Death Valley to Mt. Whitney, road race. The strain on the cardiovascular system to provide adequate blood flow to working skeletal muscle for oxygen delivery and blood flow to the skin for heat dissipation under these conditions is enormous. Further imagine that these competitors are sweating profusely and must sustain blood flow to muscle and skin when dehydrated, which reduces circulating blood volume. Under these conditions, some runners will experience dizziness, headache, confusion, and an altered gait. This is the onset of heatstroke. Recall the staggering finish of Gabriella Andersen Scheiss during the first women's Olympic marathon in 1984. This type of heatstroke is known medically as "exertion-induced heatstroke." Both forms of this multiorgan system injury can be fatal, although exertional heat stroke usually is associated with greater morbidity and mortality.

Heatstroke remains a serious problem throughout the world. It continues to occur in athletic events and during unexpected heat waves. However, "The major diagnostic challenge is to distinguish classical heat stroke from sepsis, and exertional heat stroke from less severe heat-induced abnormalities, such as heat exhaustion, a condition in which body temperature is about 40°C and the victim remains conscious" (Simon 1994 p.73).

Heatstroke: A Burning Brain or a Leaky Gut?

Is heatstroke the result of a "burning brain," that is, the direct effect of heat on brain tissue, causing lesions and denaturation of protein? Or is heat stroke the result of primary damage to body tissues that ultimately affects the brain? Historically, it has been attributed to either a central or a peripheral event. In the former case, Malamud et al. (1946) hypothesized that heat had a direct effect on the hypothalamus that led to thermoregulatory failure, the cessation of sweating, inadequate peripheral circulation, and shock (a "burning brain"). Thus, the brain could be the primary target of heat injury. Adolph and Fulton (1923–119-24) were the first investigators to attribute heatstroke to a peripheral dysfunction, that is, circulatory failure leading to shock. Acute circulatory failure was shown to precede death in more than 80 percent of 100 heatstroke cases (Austin and Berry 1956). Based on this knowledge, the critical question seems to be What causes systemic hypotension during heat stress? Is the critical target organ the gut or the brain?

Does the Fire Start in the Gut?

Research has established a mechanistic link between the splanchnic circulation and the etiology of heatstroke (figure 7.1). The splanchnic circulation—the stomach, spleen, pancreas, intestine, and liver—contains about 20 percent of the total blood volume and receives a similar portion of the cardiac output. Blood flow resistance offered by the splanchnic vasculature represents a major portion of the total peripheral resistance, and this bed is intimately involved in determining normal systemic blood pressure.

During heat exposure, blood flow through the splanchnic vascular bed decreases as flow to the skin increases to facilitate heat transfer from the core to the skin. Kielblock et al. (1982) were the first to suggest that circulatory collapse with heatstroke was attributable to splanchnic vasodilation. Kregel et al. (1988) demonstrated that splanchnic blood flow decreases throughout the early stages of heating, then increases sharply

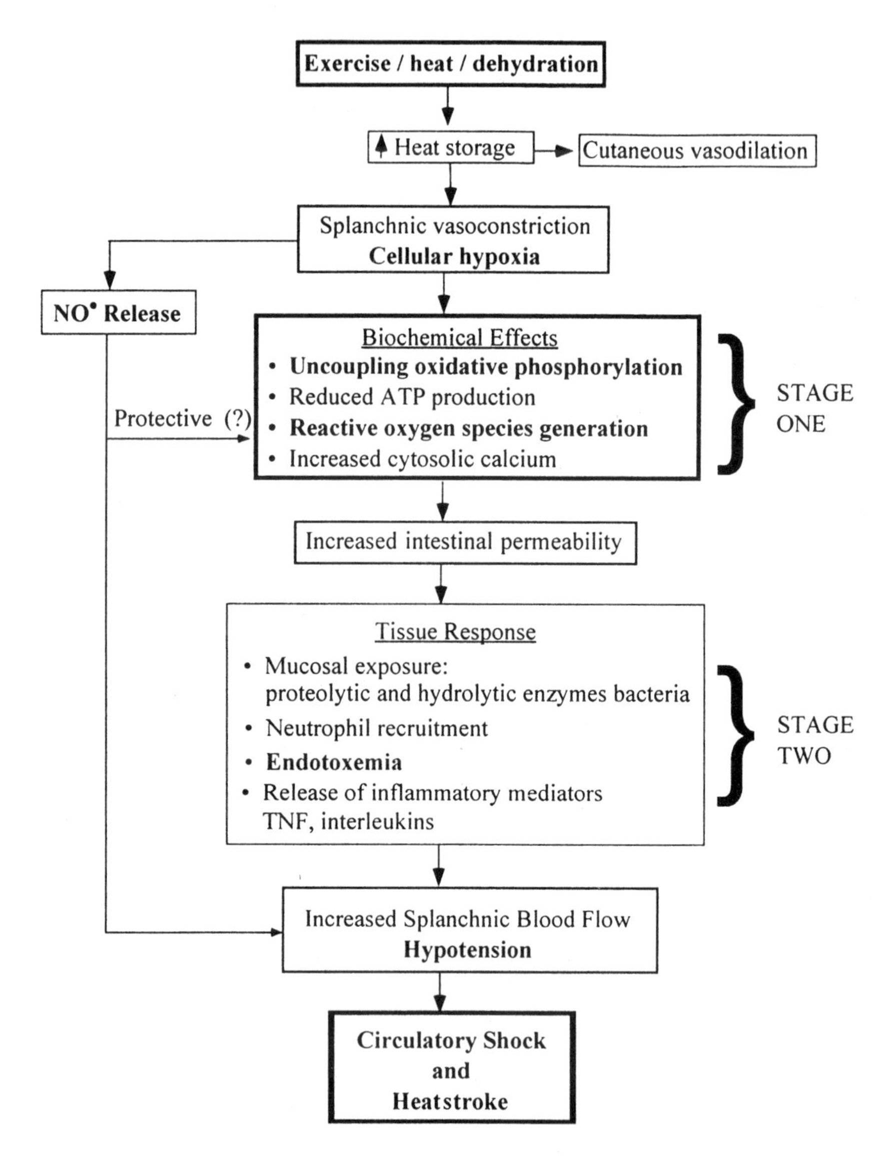

Figure 7.1
Flow diagram illustrating a hypothesis to explain the etiology of heatstroke.

10–15 min before a precipitous fall in mean arterial blood pressure. Blood flow to the kidney and to the tail of a rat (its primary heat-dissipating organ) increases during the prodromal period of heatstroke. Thus, a selective loss in splanchnic vascular resistance during heat exposure may trigger the cascade of events leading to heatstroke. The mechanism(s) responsible for this decline in splanchnic vascular resistance are unknown. Splanchnic vasodilation in the hyperthermic animal is not the result of a decrease in sympathetic nerve activity or circulating catecholamines (Gisolfi et al. 1991), or a direct effect on the vascular contractile machinery (Kregel and Gisolfi 1990).

Death from heatstroke has been observed when core temperature rises to 41–43°C. As a consequence of this high temperature, a decrease in tissue blood flow (ischemia) occurs. This can be followed by an increase in potassium concentration in the blood (hyperkalemia) and an increase in lactic acid production (acidosis). The loss of ion gradients of the cell and the lack of energy production produce tissue damage and death. The brain, liver, and small intestine, which are sensitive to ischemia and reperfusion injury, are the prime targets of tissue injury in heatstroke. When oxygen transport to the liver was reduced by lowering arterial oxygen content (normal flow), then compared with ischemia (normal arterial oxygen content), it was ischemia that produced liver damage (Tashkim 1972). On the other hand, heat stress (increasing core temperature from 37°C to 41.5°C in rats) increased the content of the hypoxic cell marker ^{3}H-misonidazole by 80 percent in the liver and by 29 percent in the small intestine (Hall et al. 1999). Furthermore, the small intestine is slow in adapting to changes in oxygen delivery, which suggests that it is sensitive to a reduction in oxygen tension (hypoxia).

It can be envisaged that the combined effects of heat, increased metabolism, reduced blood flow, and hypoxia reduce the barrier function of the gut. This barrier consists of tightly packed epithelial cells, intestinal secretions, and specialized immune cells. When it breaks down, the millions of bacteria normally present in the gut that facilitate digestion can gain access to the blood. This causes bacteremia and endotoxemia (see chapter 8). Systemic endotoxemia has been observed in heatstroke victims (Graber et al. 1971; Coridis et al. 1972) who were ultramarathon runners who collapsed during competition (Brock-Utne et al. 1988) and following strenuous exercise (Bosenberg et al. 1988). The time course of this endotoxemia has been determined in the heat-stressed monkey (Gathiram et al. 1987a), and prevented by administering a prophylactic dose of corticosteroid (Gathiram et al. 1988). A follow-up study (Gathiram et al. 1987b) showed that prophylactic corticosteroid increased survival of experimental heatstroke in primates, possibly by suppressing plasma lipopolysaccharide concentration. Thus, an increase in gut permeability may be a key factor leading to endotoxemia and hypotension.

What is the link between increased intestinal permeability (which is considered to be an increase in permeation through tight junctions) and the translocation of endotoxin (a much larger molecule), which is presumed to be a transcellular event? Figure 7.1 illustrates the two stages that characterize the progressive effects of hyperthermia (Somasundaram et al. 1995). In stage one, heat produces tissue ischemia, which in turn produces biochemical changes that uncouple oxidative phosphorylation, deplete ATP, and increase Ca^{+2} efflux from both mitochondria and endoplasmic reticulum. These events lead to increased cytosolic Ca^{+2} concentration (Carafoli 1987), generation of reactive oxygen species (superoxide, hydrogen peroxide, hydroxyl radical), alterations in cellular osmotic balance, and loss of tight-junction control, thereby producing increased intestinal permeability.

The increase in intestinal permeability is proposed as the central mechanism in the transition from the biochemical (ultrastructural) changes in stage one to the tissue reaction (macroscopic) changes observed in stage two. When the tight junctions open, their maximal channel size is too small to permit passage of endotoxin, but it does allow passage of dietary antigens and chemotactic oligopeptides (Ferry et al. 1989; Helton 1994). The latter stimulate intraepithelial lymphocytes to secrete interferon-γ. These lymphocytes are wedged between epithelial cells beneath the tight junctions. Interferon-γ opens tight junctions (Madara and Stafford 1989), and activates macrophages and neutrophils to release oxygen radicals and the immunosuppressive peptides (Helton 1994). Thus, increasing intestinal permeability by opening tight junctions can initiate immunologic and inflammatory events that can alter gut structure and function (stage two).

Thus, there is considerable evidence that a leaky gut can produce the systemic effects so frequently observed in heatstroke patients. These data also reveal why it can be so difficult to distinguish between the symptoms of heatstroke and those of sepsis. However, let us now turn to the brain, because the hypotension presumably caused by the progressive effects of a leaky gut, endotoxemia, and release of cytokines leads to cerebral ischemia, the release of toxic neurochemicals, and neuronal damage. These effects in the brain may participate in the initiation and/or exacerbation of the peripheral events outlined above.

Or Is It in the Brain?

The brain is especially vulnerable to hyperthermia-induced dysfunction (Brinnel et al. 1987). In fact, the reaction of the brain to thermal injury is complex and multifactorial. The sequence of events probably begins with general circulatory failure produced by the events in the gut. Moreover, the direct effects of heat on brain tissue may play an important role in the final pathological outcome of this process. It is a well known that during ischemia, or any other pathological insult to the brain, an increase in temperature can potentiate neural damage. But before starting with the brain, let's see what the neurological symptoms of heat stroke are, and then speculate on how the brain is involved.

The presentation of heatstroke usually is acute. Clinically, the loss of consciousness is a constant feature. However, about 20 percent of the patients have prodromal symptoms (those which precede the loss of consciousness) lasting minutes to hours and including dizziness, weakness, and nausea. A clinical picture of heat stroke can be seen in table 7.2.

Symptoms besides those listed above include seizures, stupor, delirium, irritability, and aggressiveness. In a smaller percentage of patients, symptoms include fecal incontinence, flaccidity, and hemiplegia. Prominent symptoms that could persist after recovery are cerebellar deficits, such as dysarthria and ataxia. Other symptoms that could remain after recovery are hemiparesis, aphasia, and mental deficiency. Thus, the general symptomatology, apart from the cardinal feature—the loss of consciousness—is variable and depends on individual characteristics.

Malamud (Malamud et al. 1946) reported in a clinical study of 125 fatal heatstroke cases that the most prominent pathological findings in the brain during autopsy were general edema and microhemorrhages. He reported swollen neurons and dendrites in the cerebral cortex (particu-

Table 7.2
Clinical features of heatstroke

Symptoms	Patients with Symptoms (%)
Coma	100
Convulsions	72
Confusion and/or agitation	100
Hypotension (syst. below 90)	35
Dry skin	26
Rectal temperature 41°C	55
Vomiting	71
Diarrhea	44

Adapted from Shibolet 1962.

larly the frontal cortex) and basal ganglia. But the most dramatic damage in the central nervous system appeared in the cerebellum. This consisted of edema of the Purkinje layer and a reduction in the population of Purkinje cells. The Purkinje cells remaining were swollen, pyknotic, or disintegrated. Moreover, there was a notable increase in glial mass. In contrast to these pathological findings, and despite a careful pathological analysis, no significant alterations were observed in the hypothalamus. Nor were there any demonstrable changes in the midbrain, pons, medulla, or spinal cord; mild damage and slight gliosis were found in cells of the inferior olivary nuclei and in the reticular formation. Unfortunately, the above findings were observed in fatal cases and could have developed during many hours preceding death. Therefore, they do not necessarily indicate the initial cascade of events leading to the damage reported.

Research in experimental animals has provided insight into the biochemistry, physiology, and histological changes occurring in the brain at the onset of heatstroke. Experimentally, in the rat, during the onset of heatstroke (time at which mean arterial blood pressure begins to decline from its peak level during exposure to an ambient temperature of 42°C), subjects display arterial hypotension, intracranial hypertension, decreased cerebral perfusion, degeneration of neurons with replacement by microglia proliferation, and neuronal loss (Kao and Linn 1996). Also, evidence indicating that a deterioration of the blood-brain barrier (BBB) occurs in many parts of the brain has accumulated. This deterioration can lead to increased movement of various hormones, ions, some serum proteins, and other substances from the vascular compartment into the brain. These substances could play a role in producing edema (Olsson et al. 1995; Sharma et al. 1997). In fact, some of these substances have been shown to mediate an increase in BBB permeability.

Despite these events, the primary cause of brain damage during heatstroke has been shown to be ischemia or hypoxia of the brain following the development of systemic hypotension, the result of which is the massive release of various neurotransmitters in the brain. As a consequence, an increase in free cytosolic Ca^{2+}, free radical production, and protein denaturation could be playing critical pathological roles. The direct effect of heat on the brain potentiates this ischemic damage (Ginsberg et al. 1992). Most probably the combined effects of hypotension and its consequent reduction of cerebral blood flow, together with the direct effect of heat on brain tissue, deplete ATP in many neurons of the brain.

It seems inconceivable that substances which play a physiological role in transmitting information between neurons in the brain (neurotransmitters), can become neurotoxins when accumulated in excess in the extracellular space. That is the case for at least glutamate and dopamine, which probably are two of the most important neurotransmitters known today and play crucial roles in circuits of the brain coding for behaviors such as emotion, motivation, motor behavior, and cognitive functions. In particular, glutamate has been considered one of the most important neurotransmitters in the cerebral cortex, providing rapid communication of information among neurons. Other neurotransmitters, including serotonin and noradrenaline also have been shown to accumulate in the extracellular space under heatstroke conditions.

When a Good Thing Turns Bad: Glutamate in the Brain

Glutamate is synthesized in many neurons in the brain, mainly in the cerebral cortex. Once it is released from the presynaptic terminals, it acts on postsynaptic neurons through different types of receptors: NMDA, AMPA, and metabotropic. Among these, the NMDA receptors are the most relevant.

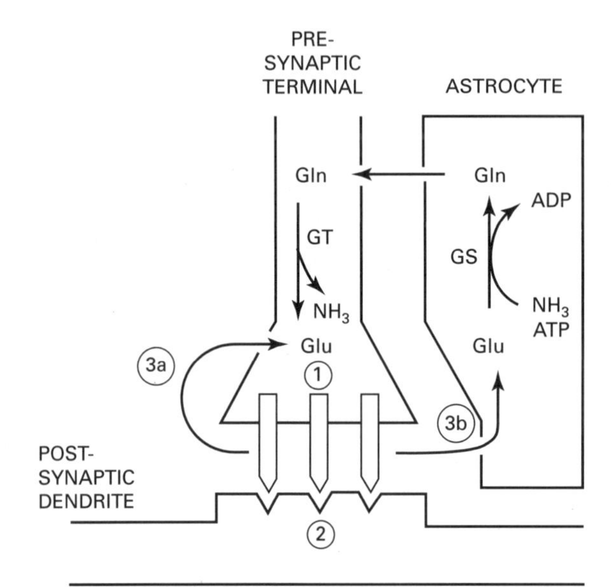

Step	Effect of Hypoxia	Potential as Site for Stroke Treatment
1. Release	Increased	+
2. Receptor binding	?	+ +
3. Neuronal (a) and glial (b) reuptake	Decreased	?

+ = possible, ++ = more likely site for stroke treatment.

Figure 7.2
A schematic diagram showing glutamergic synapse, including the presynaptic nerve terminal and postsynaptic dendrite containing glutamate (Glu) receptors. Also shown is an astrocyte. The numbers correspond to Glu release (1), binding to postsynaptic receptors (2), and reuptake by the presynaptic terminal (3a) and the astrocyte (3b). Interconversion between Glu and glutamine within the synaptic terminal and astrocyte also are shown. The table illustrates the pathophysiological steps (circled numbers in the figure) involved in hypoxic-ischemic injury. ATP, adenosine triphosphate; ADP, adenosine diphosphate; Gln, glutamine; GS, glutamine synthase; GT, glutaminase. (From Rothman and Olney 1986.)

In the presynaptic machinery, glutamate is synthesized and released by relatively well-known processes. But perhaps most relevant for our purpose is to know that glutamate, in order to transmit information rapidly, must be removed from the synaptic cleft very quickly once it has been released. Reuptake transporters located in the presynaptic terminals and astrocytes are responsible for maintaining a basal physiological concentration of glutamate in the synaptic cleft. These reuptake transporters are dependent upon normal concentrations of sodium and potassium ions inside and outside the cell, which in turn are dependent upon ion pumps that require ATP (figure 7.2).

Under hypoxic conditions, a lack of ATP synthesis leads to the reversal of ion concentrations across the presynaptic membrane, which in turn results in the release of excessive quantities of calcium-independent glutamate from the presynaptic terminals through the reverse action of the high affinity transporter for glutamate (GLU). This excessive amount of GLU in the extracellular space will permanently open NMDA receptors, through which calcium ions will enter the postsynaptic cell. The subsequent rise in postsynaptic cytosolic calcium concentration results in damage and eventual death of the cell (figures 7.3 and 7.4).

Dopamine and serotonin have been implicated in the pathophysiology of heatstroke. During heatstroke, cerebral ischemia increases extracellular concentrations of dopamine, serotonin, and norepinephrine in the hypothalamus, corpus striatum, and other areas of the brain (Lin 1997). Rats with depleted stores of dopamine in the striatum, following the destruction of dopamine cell bodies with a neurotoxin, showed less arterial hypotension, reduced intracranial hypertension, less ischemic damage in the striatum, and prolonged survival time under hyperthermic conditions. These results suggested that increased amounts of dopamine in the extracellular space could lead to the auto-oxidation of dopamine into toxic quinones and an increase in the production of cytotoxic free radicals. Similar conclusions were drawn by others who investigated the effects of ischemia in the nigrostriatal dopamine pathway.

Today, evidence is accumulating to show that it is not the release of a single neurotransmitter that produces a response in the brain, but the interaction of two or more neurotransmitters that forms the basis of all biochemical and pathological events in the central nervous system. This is

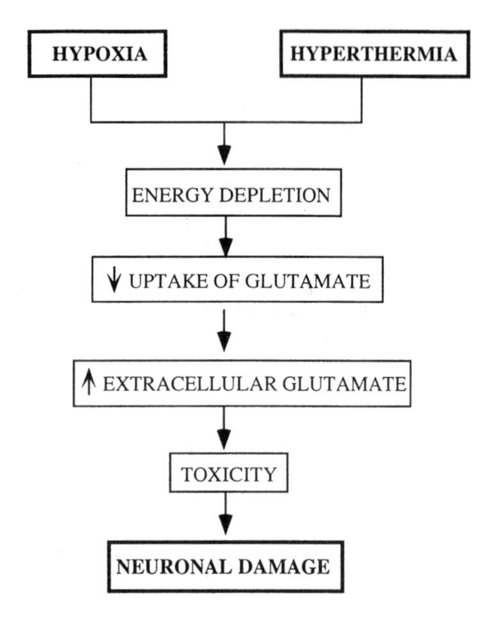

Figure 7.3
Hypoxia and hyperthermia share a common mechanism of neuronal injury mediated by glutamate toxicity.

especially the case for glutamate and dopamine in the neostriatum. The circuitry involved in this interaction, which also includes γ-aminobutyric acid (GABA) and acetylcholine (Ach), is being deciphered (Segovia et al. 1997).

As mentioned above, serotonin participates in the neuronal damage produced by ischemia. Destruction of serotonergic neurons significantly attenuates the neuronal damage associated with heatstroke, thus supporting a role for serotonin in the ischemic/hypoxic damage produced during heatstroke. Other neurotransmitters are involved in the pathogenesis of heat stroke in the brain. For instance, opioid neurotransmitters (particularly dymorphins) seem to play an important role in hyperthermic brain injury because pretreatment with naloxone or naltrexone, two opioid receptor blockers, exert a significant neuroprotective effect against the deleterious effects of heatstroke (Sharma et al. 1997). Also, the free radical nitric oxide has been implicated in hyperthermic syndromes. Increased immunoreactivity to nitric oxide synthase (NOS) in different areas of the

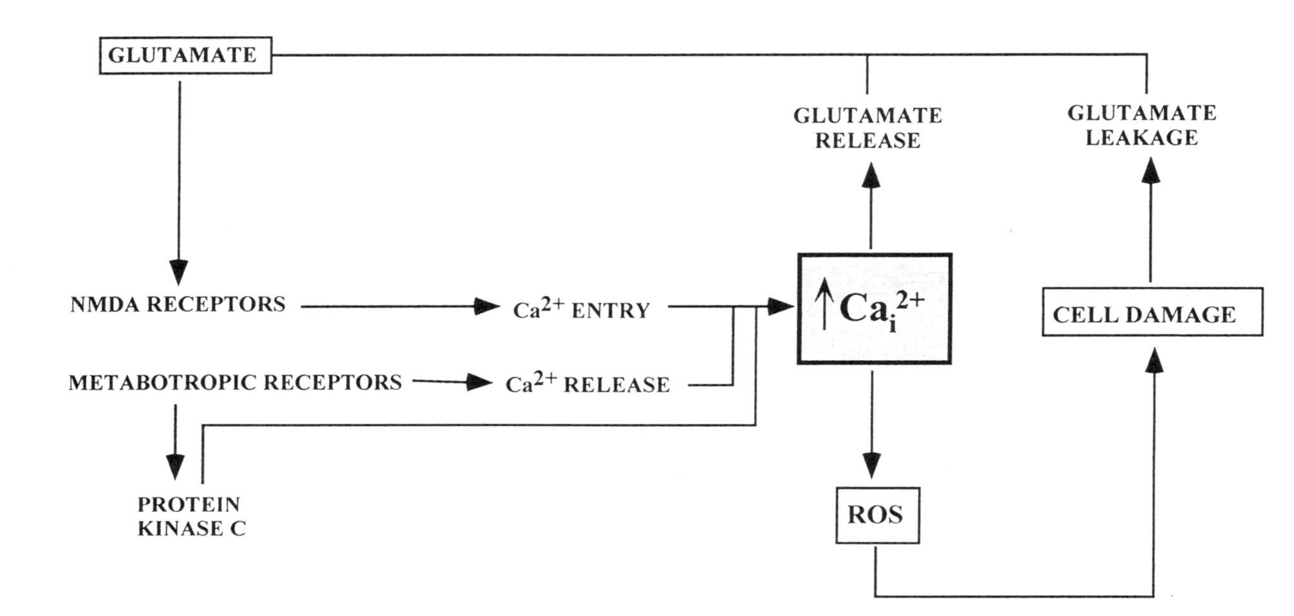

Figure 7.4
Simplified diagram showing how calcium could mediate GLU-induced neuronal degeneration. GLU acts mainly on NMDA receptors to increase cytosolic free calcium ($Ca._{}^{2+}$). $Ca.^{2+}$ in concert with diacetylglycerol, activates protein kinase C to increase neuronal excitability, further increase cytsolic $Ca.^{2+}$, and to generate reactive oxgen species (ROS) leading to death of the cell. GLU release from synaptic terminals or leakage from ruptured neurons contributes to additional injury.

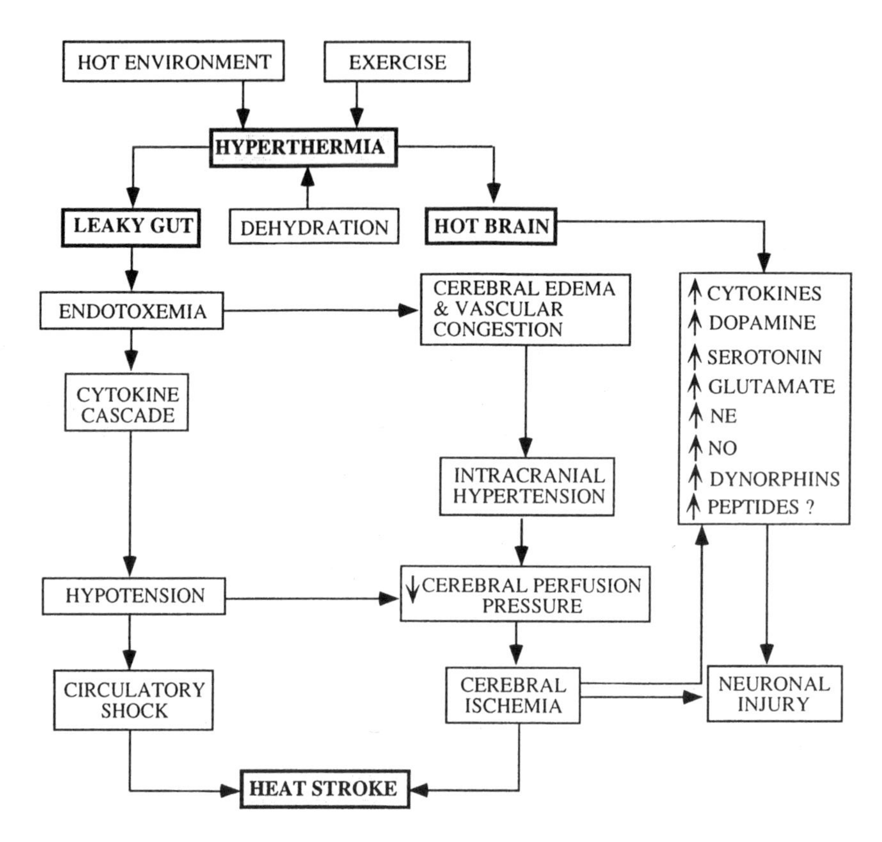

Figure 7.5
Speculative flow diagram illustrating how the combined effects of a leaky gut and hot brain can produce heatstroke.

brain was noted in rats subjected to heat stress. Moreover, pretreatment with the antioxidant compound H–290/51 showed a significant attenuation of NOS in response to heat stress, and the signs of neuronal damage and edema were less pronounced.

In summary, heatstroke, like septic syndrome, is associated with multiorgan dysfunction, endotoxemia, and the release of cytokines. These complications are positively correlated with increased intestinal permeability (Ziegler et al. 1988; LeVoyer et al. 1992). Moreover, when endotoxin is administered to humans, intestinal permeability increases (O'Dwyer et al. 1988). It is hypothesized that a "burning brain" (cerebral cell damage and neuronal loss) during heatstroke is a consequence of a decrease in blood pressure following increased intestinal permeability. Figure 7.5 represents a working hypothesis showing the events that occur in the body and in the brain during hyperthermia and their interaction which could eventually lead to heatstroke.

Heatstroke Survivors: Are They More Susceptible to Thermal Injury?

In the 1950s and 1960s, there were numerous cases of heatstroke among the Bantu laborers who worked in the gold mines of South Africa. It was only after instituting the policy that everyone who worked in the mines had to be heat-acclimatized that the incidence of heatstroke markedly declined. As a result of the heat acclimatization process, individuals who were heat-intolerant were identified and prevented from working under such hot, humid conditions. Heat intolerance has been defined as the inability to acclimatize to exercise in a hot environment (Strydom 1980). In normal healthy humans, heat acclimatization occurs in 7–10 days, but the exact differences in the physiological responses among heat-tolerant and heat-intolerant individuals have not been defined. The factors underlying heat intolerance are outlined in table 7.3. Note that prior heatstroke is among the factors listed.

In a relatively well controlled study by Israeli investigators, 9 young men who had suffered a documented heatstroke 2 to 5 years earlier were found to be intolerant of the heat, based on their physiological responses to a single heat tolerance test, compared with control subjects matched for age, height, weight, surface area, and VO_2 max (Shapiro et al. 1979). When they were asked to exercise at 35 percent VO_2 max for 3 h in a 40°C environment, their core temperature rose to 39.6°C in 1 to 2 h, heart rate rose to 160 bpm, and the subjects stopped. Control subjects were able to achieve thermal equilibrium at 38.8°C and completed the 3 h exposure. Sweat rate was the same in both groups. The investigators concluded that the heatstroke victims had difficulty transferring heat to the skin. Whether or not these patients could acclimatize to exercise in the heat as a result of repeated exposure was not determined. Thus, by the definition above, it is unclear if they were truly heat-intolerant.

Table 7.3

Factors underlying heat tolerance
a. General Factors
Dehydration
Unfitness
Heavy clothing
No acclimation
Previous heatstroke
Old age
Fatigue
b. Diseases
Cardiovascular diseases
Infectious diseases
Psychiatric disturbances or diseases
Parkinson disease
Ectodermal dysplasia
"Chronic idiopathic anhidrosis"
Diabetes
Extensive burn scars
Hyperthyroidism
Cystic fibrosis
Scleroderma
d. Drugs
Drug consumers

In a study by Armstrong et al. (1990), 9 of 10 prior heatstroke victims were able to acclimatize to dry heat within 2 months after their heatstroke. One patient was unable to acclimatize at this time and was classified as being heat-intolerant, but he successfully acclimatized to exercise in the heat 11.5 months after his heatstroke. The difference between the 9 patients who successfully acclimatized within 2 months of their heatstroke and the one person who required 11.5 months to display the appropriate physiological responses to the heat probably is related to the time lapse between the onset of the injury and treatment, which is the primary factor that determines survival (Shibolet et al. 1976).

The Challenge: Only 4°C Separates Life from Death!

The normal resting core temperature of humans is about 37°C (98.6°F), and the core temperature threshold for heat stroke is only 40.6°C (105.1°F) (Leithead and Lind 1964). Thus, only 4°C separates life from potential death. Does the brain have a limited capacity to tolerate heat? If so, why? What factors might be responsible for such limited capacity? Tolerance for a high core temperature varies among and within species (Adolph 1947; Hubbard et al. 1977). For example, two species of antelope, Grant's gazelle and the oryx, can survive desert life, in part, by allowing their core temperature to rise above 46°C (Taylor 1970). And human marathon runners have been shown to tolerate core temperatures of 40–42°C. Such tolerance may be an advantage when one is exposed to high thermal loads, and contribute to survival. In this context, a crucial question for humans is Can the 4°C range over which we normally regulate core temperature be expanded?

Can the 4°C Thermoregulatory Range Be Expanded? Training and Acclimatization

Physical conditioning and heat acclimatization have been shown to improve physiological function. Physical training increases the size of the heart and its ability to pump blood, which in turn leads to an increase in performance. Training also increases the number and size of mitochondria (factories for generating ATP), and the number of capillaries in the heart and skeletal muscle. The latter effect improves myocardial blood flow and helps to protect against heart attacks. Heat acclimatization increases the ambient temperature range over which body temperature is regulated, and physical training increases thermal tolerance. However, can a highly trained and/or heat-acclimatized individual tolerate a higher core body temperature, or are highly trained athletes more susceptible to thermal injury because of their greater capacity to generate metabolic heat? Exciting studies indicate that physical training and/or heat acclimatization may allow tolerance to higher core body temperatures.

Exertional heatstroke often occurs as a result of exercise at high rates of metabolic heat production in warm environments. Table 7.4 shows the

Table 7.4
Observed and derived results on the first competitors finishing the Whitney Marathon

Order of Finishing	1	2	3	4
	Observations			
Time (min)	158	162	164	166
Rectal temperature (°C)	41.1		40.5	40.2
Weight loss (kg)	5.23	3.12	2.73	4.66
	Derived results			
Volume of sweat (liters)	4.76	2.79	2.42	4.24
Sweat rate (liters, hr^{-1})	1.81	1.03	0.89	1.53
Total fluid loss (liters)	5.10	3.07	2.69	4.56
Fluid loss (as percentage of weight)	6.9	5.1	4.7	6.6

Modified from Pugh et al. 1967.

observed and derived responses of the first four finishers of a marathon (Pugh et al. 1967). The winner had a core temperature of 41.1°C. Core temperature of the 2nd-place finisher was not taken, and the 3rd- and 4th-place finishers had rectal temperatures above 40°C. The investigators concluded that tolerance of a high body temperature was a necessary condition for success in this event. Another study revealed that a marathon runner maintained a core temperature between 41.6° and 41.9°C for the last 44 min of a race without sustaining thermal injury (Maron et al. 1977).

In a study designed to determine if training in a cool (23°C) environment would alter the potential for mortality or tissue damage during a work-heat tolerance test to exhaustion, rats were trained on a motorized treadmill for 6 weeks. Tissue damage and percent mortality were the same in the trained and control sedentary animals, but the trained survivors (a) continued the test 44 percent longer, (b) performed significantly more work, and (c) sustained a 120 percent larger thermal load (product of time and colonic temperature above 40°C) than sedentary survivors. Mortality first occurred at a core temperature range of 40.6–41.0°C in the sedentary animals, compared with a range of 41.6–42.0°C in trained animals (figure 7.6). Thus, training enabled rats to run longer in the heat,

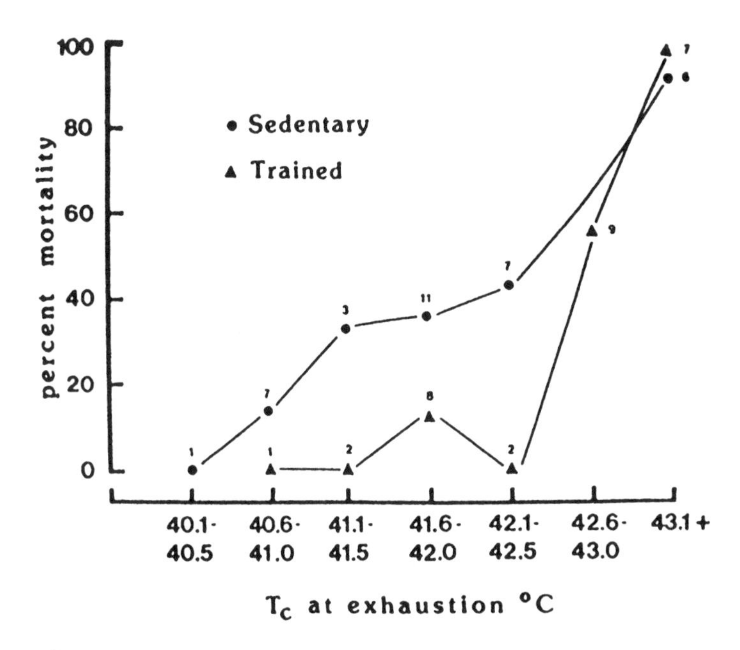

Figure 7.6
Percent mortality during work-heat tolerance test in trained and sedentary animals. Numbers above data points indicate the number of animals in each group. (From Fruth and Gisolfi 1983.)

sustain greater thermal loads, and be less susceptible to work-induced thermal fatality than sedentary animals.

To elucidate the mechanism underlying the greater thermal tolerance of highly trained endurance athletes and endurance-trained animals, Sakurada and Hales (Sakurada 1997) determined the effect of intravenous indomethacin on core temperature of physically fit and sedentary sheep exposed to heat (42°C). Indomethacin blocks prostaglandin pathways involved in fever induced by endotoxin. The observation that indomethacin reduced the rate of rise of core temperature in the sedentary sheep, but not the physically fit sheep, indicates that endotoxin plays a role in determining heat tolerance. Presumably, the greater heat tolerance of physically fit animals is related to their greater splanchnic blood flow, which permits better maintenance of intestinal barrier function. This hypothesis is supported by the greater splanchnic blood flow in heat-acclimatized animals (Shochina et al. 1996).

Can heat acclimatization expand the narrow temperature zone over which core temperature is regulated? In a preliminary report, Horowitz et al. (1997) found that acclimatizing rats for one month to an environmental temperature of 34°C elevated the basal 70 kilodalton heat shock protein (see below) content in heart tissue by 140 percent. This coincided with an improved myocardial perfusion during ischemic insult. Heat shock protein (HSP) synthesis is turned on by tissue damage. Having a ready supply of HSP on hand presumably allows the animal to cope with greater thermal stress, that is, higher core body temperatures.

Neonatally Induced Thermotolerance: Unique Blend of Head Acclimatization and Acquired Thermotolerance?

After broiler cockerels (chickens) were exposed to an environmental temperature of 35.0–37.8°C for 24 h at 5 days of age, they experienced significantly lower mortality upon exposure to the same environmental conditions at 43 days of age than cockerels that did not experience neonatal heat exposure (Arjona et al. 1990). There were no differences in core temperature between neonatally heated and neonatal control animals. Thus, the heat-stressed neonates could better withstand the rise in body temperature. Moreover, the ratio of surface to core temperature was the same for the two groups, indicating that the greater thermotolerance of the heat-stressed birds was not the result of greater heat dissipation at the skin surface. Because acquired thermotolerance associated with the synthesis of HSPs is dissipated after 7 days, and because these neonates were not exposed to additional thermal stress after their 5th day of life, the improved survival observed on day 43 cannot be attributed to either heat acclimatization or HSP accumulation. HSPs were not measured in these investigations. Thus, the mechanism of this neonatally induced thermotolerance remains unknown.

A New Story: Heat Shock Proteins as Cell Thermometers, Stabilizers, and Chaperones

In organisms ranging from bacteria to man, heat exposure produces a *heat shock response*. For example, when the larvae of the fruit fly

(*Drosophila melanogaster*) were exposed to a temperature of 40.5°C, most of them died. However, if the larvae were exposed to mild heat (about 35°C) as a conditioning stimulus before exposure to the more severe heat stress, about 50 percent survived (Mitchell et al. 1979). This *acquired thermotolerance* or *thermoprotection* is attributed to the synthesis of new proteins in response to the cells being subjected to elevated temperatures (Tissieres et al. 1974). The greater the initial heat stress, the greater the thermotolerance and protection afforded to the cell. In contrast to heat acclimatization (see chapter 4), which requires 5 to 10 days to develop, thermotolerance occurs within hours of the initial exposure and lasts for 5 to 7 days (figure 7.7). The appearance and disappearance of HSP in cells parallels the acquisition and decay of thermotolerance (Li 1985); blocking HSP synthesis during the conditioning stress prevents the development of thermotolerance (Riabowol et al. 1988). Because heat was the first stressor to produce these newly discovered proteins, they were called "heat shock proteins," but it was soon discovered that other stressors, such as oxidants, alcohol, heavy metals, and microbial infection, also induced the synthesis of these proteins. Thus, the proteins are more appropriately termed "stress proteins" and the response, the "cell stress response." HSPs have been called the "thermometers of the cell" (Craig and Gross 1991).

There are several families of HSPs, and among these the 70 kilodalton (HSP70) family is the most highly conserved; 50 percent of their structure is similar between *E. coli* (bacteria) and humans, and some domains are 96 percent similar. The HSP70 family consists of (a) an abundant 73K cognate that is present in the unstressed cell but can be moderately induced by stress, and (b) a highly inducible 72K stress protein. The fact that these proteins are so strongly conserved in structure across species points to their importance in the survival of the organism. Their synthesis enhances the ability of a cell to recover from stress, but precisely how this is done is unclear.

How Does Heat Affect Cells, and What Is the Function of HSP?

Heat increases metabolic rate and, as shown in figure 7.1, can uncouple oxidative phosphorylation, leading to ATP depletion. Moreover, the metabolic products of ATP breakdown can produce reactive oxygen

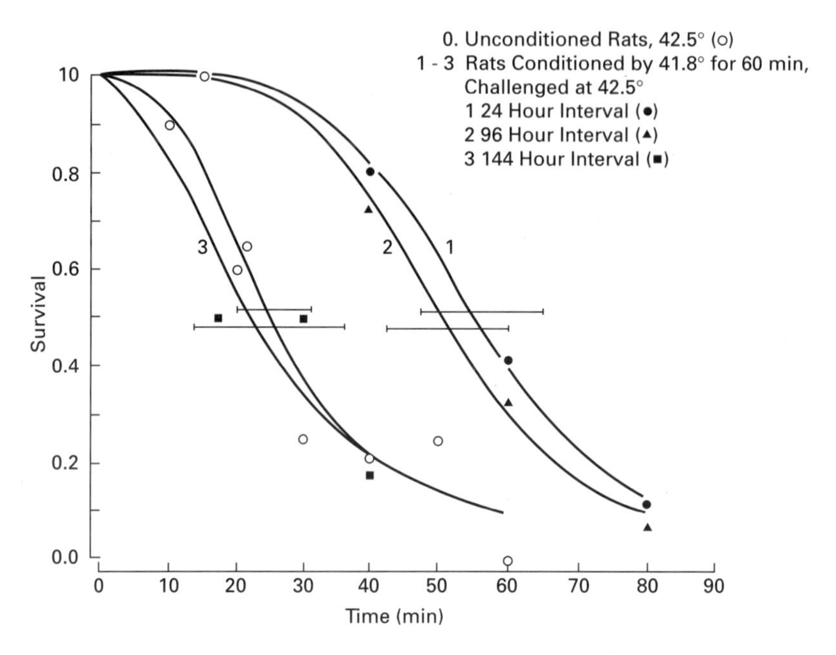

Figure 7.7
Survival after whole-body heating at an ambient temperature of 42.5°C for 24 (curve 1), 96 (curve 2), and 144 (curve 3) hours after a conditioning heat exposure at 41.8°C for 60 min. The curves represent best-fit plots determined by the logistic regression method. Horizontal bars indicate 95 percent confidence intervals for the LD_{50}. Curve 0 represents survival of previously unheated rats (control) following heating at 42.5°C. (From Weshler et al. 1984.)

species (superoxide, hydrogen peroxide, hydroxyl radical), which are detrimental to membrane lipids, cell proteins, and DNA. Abnormal proteins can induce the synthesis of HSP. One function of HSPs is to bind denatured proteins or unfolded protein fragments, and (a) serve as "foldases" to restore them to their normal tertiary structures (see figure 7.8) or (b) mark them for lysosomal degradation (Moseley 1994). Another important role of HSPs is their ability to facilitate the translocation of proteins across membranes, that is, their role as chaperones. This involves the unfolding, transmembrane transport, and refolding of proteins. Thus, acquired thermotolerance or thermoprotection may result from marking denatured proteins, from stabilizing proteins in the process of denaturation, from refolding proteins in the process of denaturation, and/or from

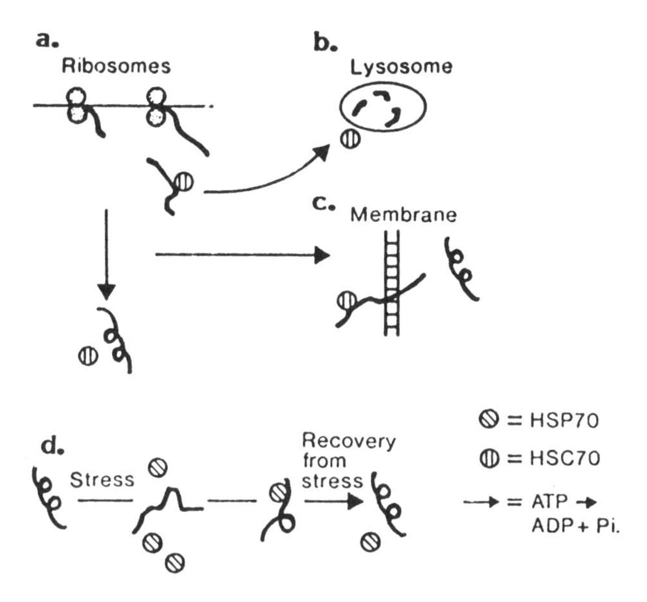

Figure 7.8
Schematic diagram illustrating the possible chaperone functions of the 70-kDa heat-shock protein (HSP70). Members of this family, when expressed constitutively (HSC70), bind to proteins to assist their proper folding and assembly (a). Wrongly folded proteins or denatured proteins are refolded or transported to lysosomes for degradation (b). HSP also assists in transporting proteins across membranes into various cellular compartments (c). Heat and other stresses denature proteins and induce HSP70 synthesis. Newly formed HSP70 binds to denatured proteins to keep them soluble and assists in their renaturation (d). Proteins are released from HSC70/HSP70 with the aid of ATP hydrolysis. (Modified from Jaattela and Wissing 1992.)

providing transport of key enzymes or structural proteins required for cell survival (Moseley 1994).

Although much of the research involving HSPs has been performed in cell culture systems, isolated organs and tissues, more recent studies on whole animals have revealed that HSPs may participate in whole-body adaptations. Results from a study of nine lizard species from various geographical regions led the investigators to postulate the following general rule for poikilothermic organisms: a direct correlation exists between the characteristic temperature of the ecological niche for a given lizard species and the amount of HSP70-like protein in its cells at normal temperature (Ulmasov et al. 1992). This postulate is supported by the incredible ability of desert ants (*Cataglyphis*) to survive body temperatures of 50°C for at least 10 min and exhibit a critical thermal maximum of 53–55°C (Gehring and Wehner 1995). This remarkable capacity for thermal stress is associated with the ability of these insects (a) to synthesize HSP at up to 45°C, compared with only 39°C for *Drosophila*, and (b) to accumulate HSP prior to heat exposure.

In addition to thermotolerance, formation of HSPs may contribute to heat adaptation of whole organisms by their effect on (a) maintenance of epithelial barrier integrity and (b) their ability to enhance endotoxin tolerance. Recall from figure 7.1 that heat stroke was hypothesized to result from an increase in intestinal permeability leading to endotoxemia and an increase in circulating cytokines. If the intestines are prophylactically sterilized, or if antiendotoxin antibodies are administered to experimental animals, they can tolerate higher core body temperatures and their survival in the heat improves (Gathiram et al. 1987a, 1987b). In an epithelial monolayer grown in culture, a reversible increase in permeability occurs with a rise in culture temperature. If these culture cells are allowed to accumulate HSP70 from a preconditioned thermal stress, the rise in permeability with hyperthermia is attenuated (Moseley et al. 1994). Thus, the accumulation of HSP70 may confer heat tolerance by helping to maintain epithelial barrier integrity. It also confers tolerance to endotoxin in animals (Ryan et al. 1992). This latter phenomenon may reflect tolerance to the direct effect of endotoxin, tolerance to cytokine exposure, or inhibition of oyotkine production by inflammatory cells. These effects of HSPs may represent a cellular mechanism associated with heat acclimatization that enables cells to continue to function at elevated temperatures, and protects tissues and organs from thermal injury.

In summary, the accumulation of HSPs under normothermic conditions may enable mammals, including humans, to expand their normal 4°C thermoregulatory zone to include core temperatures above 41°C. The remarkable thermal tolerance of some endurance athletes and the observation that heat acclimatization significantly increases resting levels of HSPs, support this postulate. In addition to the survival benefit associated with HSP accumulation through training and heat acclimatizations, numerous therapeutic benefits of HSPs are emerging (Ezzell 1995). For example, hearts from mice genetically engineered to contain human HSP70 genes recovered twice the contractile force observed in control hearts following ischemia and reperfusion. Scientists are now searching for compounds that promote HSP production. One such compound is aspirin, which acts to induce heat shock factor (HSF).

HSF is the transcription factor that regulates HSP70 gene expression as a result of binding to the regulatory elements of the HSP70 gene. The heat shock response also improves the success of organ and tissue transplantation. Elevating core body temperature of donor animals to 42.5°C for only 15 min and allowing the animal to recover for approximately 6 h improved the survival and functioning of transplanted kidneys and pancreatic islet cells. Applying a heat blanket at 45°C to the skin of rats for 30 min and allowing 6 h recovery doubled the amount of skin that survived surgical excision for subsequent grafting. However, not all is positive. Some HSPs may aid the metastasis of cancer. Women with breast tumors that produce HSP27 had shorter periods of disease-free survival than women with lower levels of HSP27. This overproduction of HSP27 in breast cancer metastasis may help to explain why some tumors become resistant to chemotherapy after a while and why some tumors become resistant to hyperthermic therapy. Whether beneficial or detrimental, there is much more to learn about these unique proteins.

Running in the Heat: What Makes Us Stop?

An English physician once commented that the only real threat to life during exercise is the possibility of suffering a fatal heatstroke. However, when a fit but untrained and unacclimatized healthy man or woman engages in hard exercise, even in a hot environment, heatstroke rarely occurs. Why? The answer is that the man or woman stops exercising before reaching heatstroke conditions. Are there specific signals telling the person to stop, thereby providing a natural defense against such injury?

The increased demand for blood flow to the skin during exercise in the heat is met in part by redistributing the cardiac output. Blood flow to splanchnic and renal vascular beds, reduced during exercise in a cool environment, is reduced further (figure 7.9). The question is whether or not blood flow to working muscle is reduced as as well, as proposed by

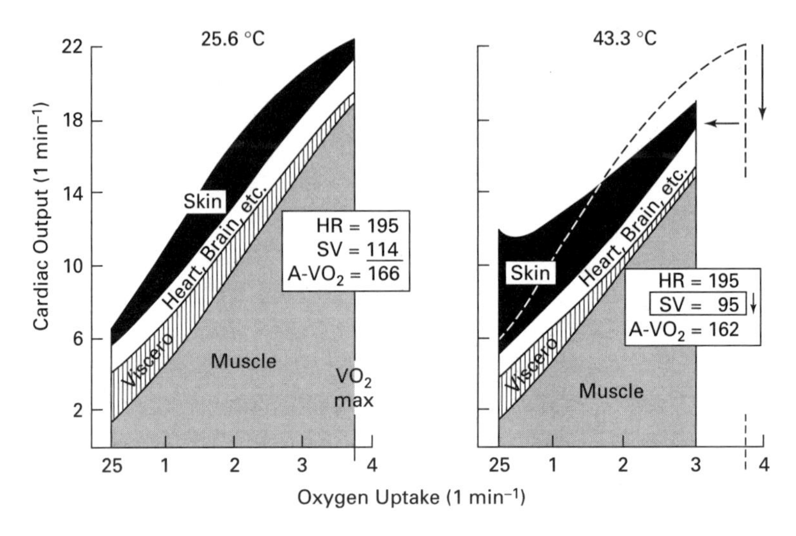

Figure 7.9
Estimated distribution of cardiac output over the range of oxygen uptake from rest to VO_2 max in neutral and hot environments. The rise in cardiac output is less than in neutral conditions (broken line in right panel) when VO_2 exceeds 2 l/min. It is suggested that skin blood flow is elevated during heat stress, at the expense of reducing muscle blood flow. (From Rowell 1986.)

Rowell (1986), and whether such a reduction in muscle perfusion alters muscle metabolism, leading to the fatigue that limits performance in the heat.

In a study by Nielsen et al. (1990), muscle blood flow did not decline during exercise in the heat as heart rate approached maximal values and core temperature approached 40°C, but oxygen uptake rose by about 0.5 liter/min (figure 7.10). The uptake of glucose and free fatty acids in cool and warm environments was not different, indicating that energy delivery was not impeded by high muscle and core temperatures. What, then, is the cause of exhaustion during exercise in the heat? Nielsen et al. (1990) concluded, as have others (Bruck and Olschewski 1987), that this exhaustion stems from the effects of high temperature on CNS function, that is, motor center function and/or motivation for motor performance.

In a subsequent study by Nielsen et al. (1993) designed to investigate the circulatory and thermoregulatory adaptations with heat acclimatization, it was again concluded that high core temperature per se, and not

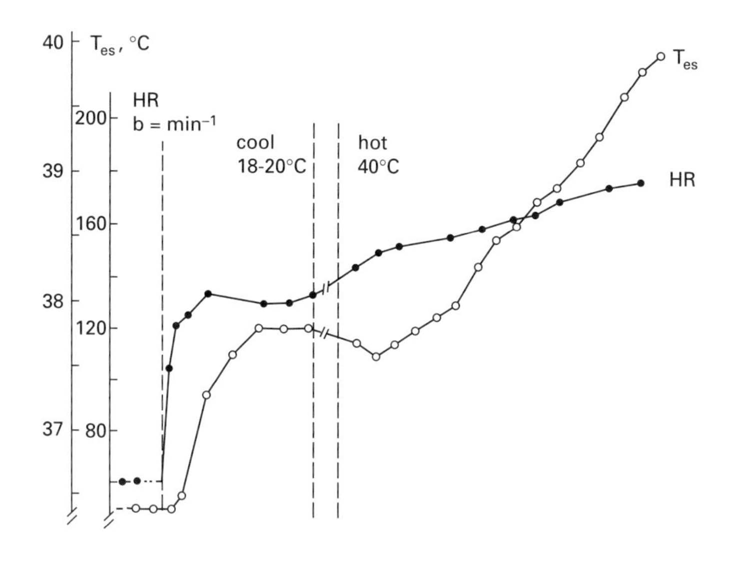

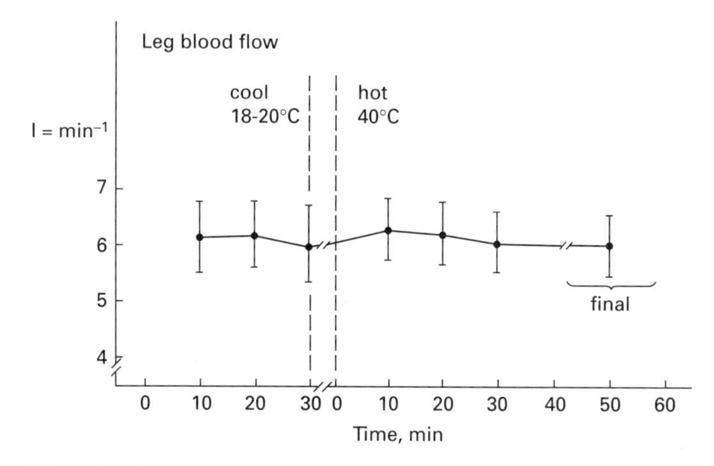

Figure 7.10
Esophageal temperature (T_{es}), heart rate (HR), and leg blood flow measured as femoral vein flow during 30 min of walking in cool environment immediately followed by up to 60 min of walking in a hot environment. T_{es} and HR are for one representative subject; values for blood flow are means ± SE for 7 subjects. (Modified from Nielsen et al. 1990.)

circulatory failure, was the critical factor leading to exhaustion during exercise in the heat. In this study, subjects exercised to exhaustion at 50 percent VO_2 max for 9–12 consecutive days at an ambient temperature of 40°C with 10 percent relative humidity (figure 7.11). They were highly motivated, and cycled to exhaustion each day without any clues as to how long they had worked. Evidence of acclimatization included increased sweating rate, and a lower rate of rise of core temperature and heart rate. Endurance time increased from 48 to 80 min. The intriguing observation was that exhaustion occurred each day at a core temperature of about 40°C, despite no reduction in cardiac output and muscle blood flow, no changes in substrate utilization or availability, and no accumulation of fatigue substances (figure 7.11). Moreover, there was no change in the ability to recruit motor units. As an index of local or central fatigue, maximal isometric force with the elbow flexor and knee extensor was measured; it was found to be the same before and after exercise.

In contrast to the data of Nielsen et al. (1990) in humans, when dogs were made hyperthermic by running with hot pack, they showed metabolic changes suggesting that a high muscle temperature limited skeletal muscle performance. Hales (1983), using sheep as an experimental model, showed a decrease in muscle blood flow during exercise in the heat. Rowell (1986) argued that the increase in cardiac output and the reductions in splanchnic and renal blood flows during exercise in the heat are insufficient to account for the estimated increase in skin blood flow in exercising humans during heat exposure (figure 7.12). If the resolution of the thermodilution technique in the studies by Nielsen et al. (1990, 1993) is only 10 percent, and muscle blood flow during exercise is as high as 18 l/min, the measurement of muscle blood flow could be off by 1.8 l/min.

The mechanism responsible for reducing blood flow to working skeletal muscle during exercise in the heat is unclear. Increased sympathetic drive directed at skeletal muscle may be involved. Exercise in the heat produces the most marked increase in circulating catecholamines, and contracting skeletal muscle can vasoconstrict during direct sympathetic nervous stimulation (Donald et al. 1970). Intact adrenergic nerves are required for the decrease in muscle blood flow observed in resting heat-stressed sheep (Hales 1983). Thus, the issue of what limits performance in

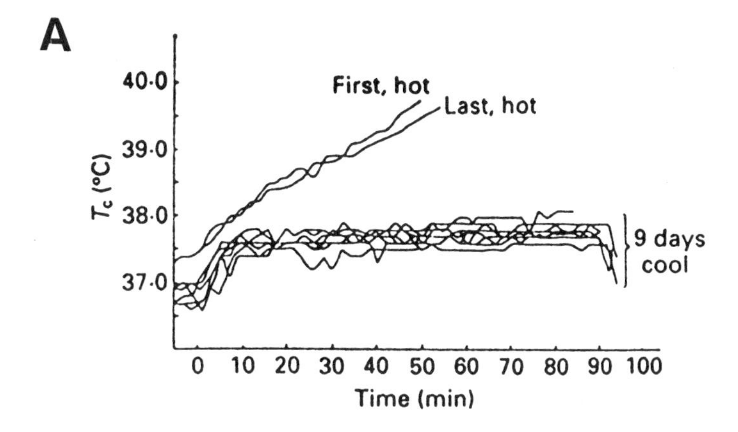

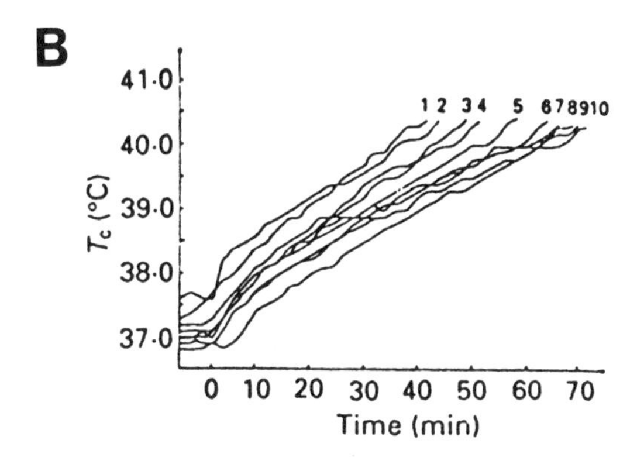

Figure 7.11
Core temperature (T_C) as a function of time. Representative data from a control subject during exercise at 40°C during the first and final experiments, and during 9 intervening rides on a cycle ergometer in a cool environment (18–20°C). (B) Representative data from an acclimatizing subject during 10 consecutive days of cycle exercise in the heat (40°C) until exhausted.

the heat remains unresolved. The data of Nielsen et al. (1990, 1993) are impressive, but the arguments put forth by Rowell (1986) for a reduction in muscle blood flow are equally compelling (figure 7.12). A decrease in performance may not be a mandatory consequence of a decrease in muscle blood flow; there is considerable room for increasing hind limb oxygen extraction in the event of a fall in muscle blood flow (Bird et al. 1981). Thus, in the end, it may be the brain that indeed limits performance in the heat. A consequence of continually overriding afferent input

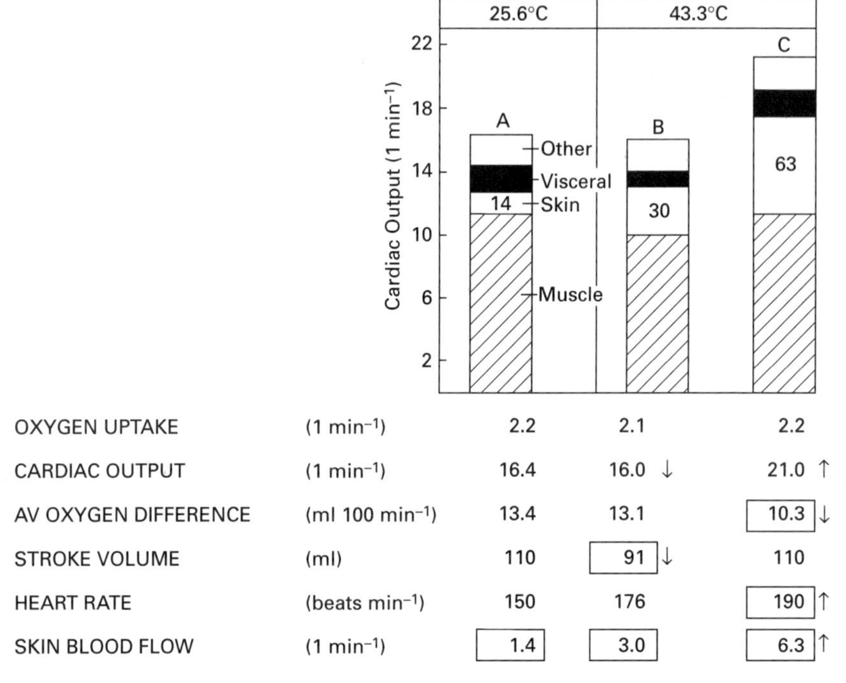

		A	B	C
OXYGEN UPTAKE	(1 min^{-1})	2.2	2.1	2.2
CARDIAC OUTPUT	(1 min^{-1})	16.4	16.0 ↓	21.0 ↑
AV OXYGEN DIFFERENCE	(ml 100 min^{-1})	13.4	13.1	10.3 ↓
STROKE VOLUME	(ml)	110	91 ↓	110
HEART RATE	(beats min^{-1})	150	176	190 ↑
SKIN BLOOD FLOW	(1 min^{-1})	1.4	3.0	6.3 ↑

Figure 7.12
Two possible means of increasing skin blood flow during moderate exercise in a hot (43.3°C) environment. Comparison is with normal responses at 25.6°C (column A). Listed below columns A and B are measured variables and estimated skin blood flow (boxes are for emphasis). Column C shows a hypothetical maximal adjustment achieved by raising heart rate to 190 bpm, while stroke volume and distribution of cardiac output remain as they were at 25.6°C. Skin blood flow could reach 6.3 l/min. This would lower systemic arteriovenous oxygen difference to 10.3 ml/100 ml. Column B shows that the actual fall in stroke volume (compensated by rise in heart rate to 176 bpm) prevented a rise in cardiac output. Thus skin blood flow could be increased to 3 l/min only by additional visceral vasoconstriction and mainly by reducing muscle blood flow to make its oxygen extraction 100 percent. The slightly lower oxygen uptake at 43.3°C (2.1 l/min) could reflect such a change. (From Rowell 1986).

to the brain that details the thermal status of the body could be exertional heat injury. Many athletes, especially in high stakes competition (the Olympics), can push themselves beyond their physiological limits and consequently suffer heat injury.

Gender Differences in Heat Tolerance

Are women less tolerant of the heat than men? Because they sweat less than men, does this mean they are more susceptible to thermal injury and heatstroke? Early comparisons of thermoregulation in men and women showed that women had lower sweat rates, higher core temperatures, and higher heart rates than men during exercise, and did not tolerate the heat as well as men (Drinkwater 1986). However, subsequent studies revealed that these differences, as well as those associated with age, may be attributed to differences in fitness.

If a group of men and women are asked to exercise at three different intensities and their steady-state core temperature is plotted as a function of exercise intensity, the lines representing the men and women are scattered (figure 7.13A). On the other hand, if each exercise intensity is converted to a percentage of VO_2 max for each subject and the steady-state core temperature is plotted as a function of each individuals' percent VO_2 max, the data converge (figure 7.13B). In other words, if all subjects work at 50 percent of their aerobic capacity, even though some will be working significantly harder than others, core temperature is the same. Back in the 1960s, when the first comparisons were made of thermal tolerance in men and women, both sexes were exercised at the same absolute workload. If the women in these studies were less fit than the men, and in all likelihood they were, they would have been exercising at a higher percentage of their aerobic capacity, which would explain their higher core temperatures and heart rates.

Age and Heat Tolerance

In July 1995, a record number of fatalities was attributed to a heat wave in the midwestern United States. The elderly, who presumably are more susceptible to heat injury, accounted for the largest number of heat-re-

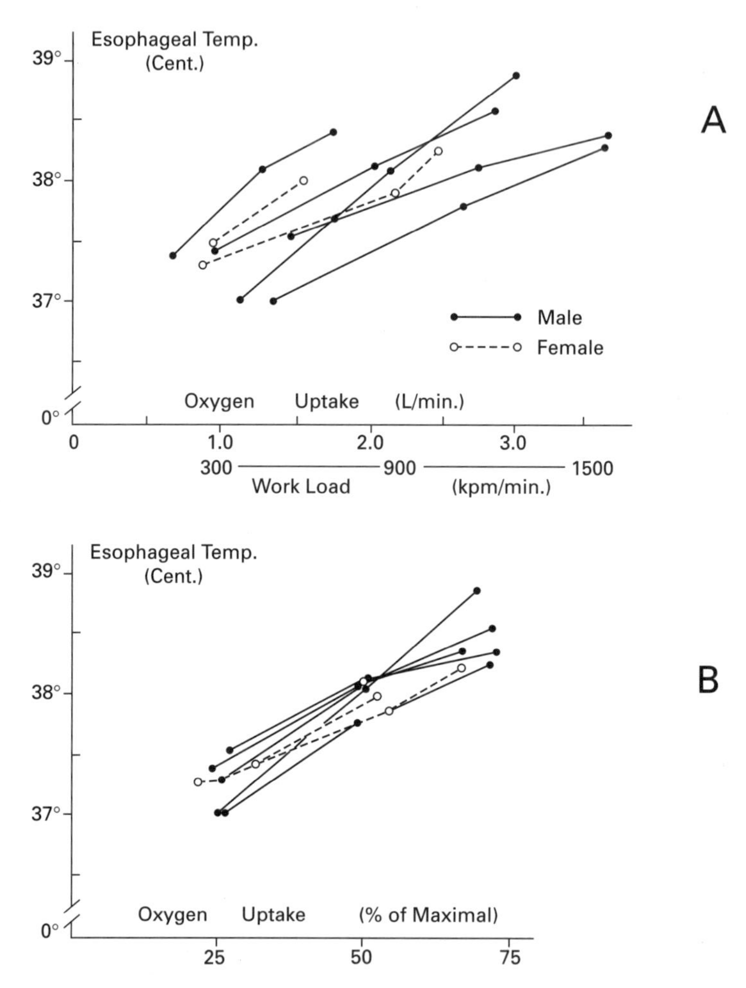

Figure 7.13
Individual values for esophageal temperature in relation to oxygen uptake (A) and as a function of percent VO_2 max (B). (From Saltin and Hermansen 1966.)

lated deaths. In humans, old age has been associated with a reduced core temperature at rest, hyperthermia in response to heat stress, reduced capacity to sense temperature extremes, and inadequate thermoregulatory responses to environmental challenge. Does aging truly result in temperature regulatory dysfunction and heat intolerance, or are these functional deficits the result of preexisting disease processes and/or differences in aerobic power, heat acclimatization, and body composition? The two primary effector mechanisms for regulating body temperature during heat challenges are skin blood flow and sweating. Is either of these mechanisms compromised with advancing age?

In a review of the thermoregulatory responses of healthy older adults at rest and during exercise, the elderly (defined as men and women over 65 years of age) did not have lower core body temperatures at rest compared with young (20–30 years of age) men and women, and had normal circadian temperature rhythms (Kenney 1997). Moreover, the number of sweat glands that responded to pharmacological stimulation was the same in older and younger individuals, but the amount of sweat produced per gland was reduced in the elderly. However, this age-related deficit in drug-activated sweating does not necessarily translate to a lower sweat rate during exercise in the heat. As illustrated in figure 7.14A, sweating rate was not different in young and old groups in a hot, humid environment, but was significantly lower in the older subjects during exercise in a hot, dry environment. This latter response was attributed to less sweat produced per gland rather than a reduction in the number of heat-activated sweat glands (figure 7.14B).

The ability of the elderly to transfer heat to the skin by increasing cutaneous blood flow is limited. Figure 7.15 shows that the threshold for vasodilation during exercise in a warm environment is not altered by age, but the rate at which blood flow increases as core body temperature rises is reduced. Moreover, the highest skin blood flow that can be attained is reduced in the elderly. The mechanism responsible for this decrease in skin blood flow is not understood. It is not the result of greater vasoconstrictor tone, because systemic blockade of α_1-receptors with prazosin had no selective effect on skin blood flow of older men (Kenney 1997).

In summary, when rigid criteria are applied to the few studies that have addressed the issue of how age affects temperature regulation, the idea

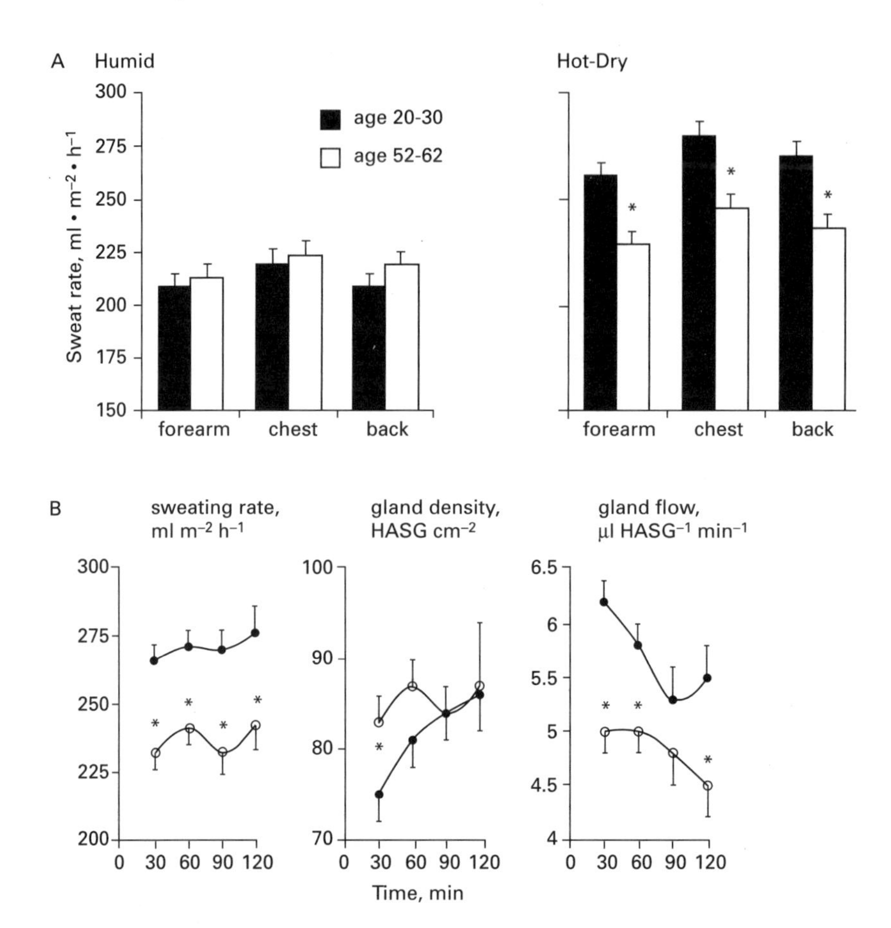

Figure 7.14
(A) Local sweating rates on three skin sites for two age groups of post-menopausal women exercising in warm-humid (37°C, 60 percent r.h.) and hot-dry (48°C, 15 percent r.h.) environments. (B) The age difference in sweating rate observed in the hot-dry environment shown in (A) is attributed to a lower sweat output per gland, not fewer activated glands. Women (n=8 per group) were matched for VO_2max and exercised at 35–40 percent VO_2 max. HASG is heat-activated sweat glands. (Modified from Kenney 1997.) *Significantly lower sweat rates compared with young subjects.

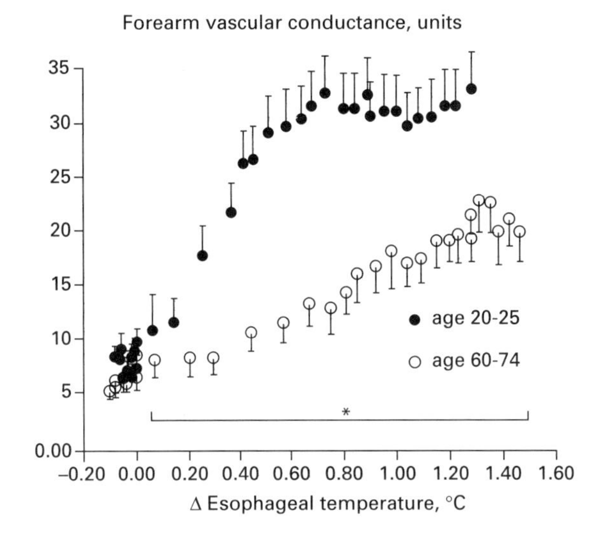

Figure 7.15
Increase in forearm vascular conductance (forearm blood flow/mean arterial pressure; representative of skin blood flow) during exercise in a warm (36°C) environment is significantly attenuated in older men (n=15 per group). (Modified from Kenney 1997.)

that heat tolerance and thermoregulatory function are compromised with advancing age is not supported. The deficits observed in heat tolerance and thermoregulatory function among the aged are most probably attributable to differences in physical fitness, heat acclimatization, and changes in body composition that accompany aging.

8

Fever, Survival, and Death

At the beginning of the twentieth century, William Osler wrote that the three great scourges of humanity were fever, famine, and war (Atkins 1991). "For much of history the word 'fever' has been used almost synonymously with disease itself as various epidemics have ravaged the civilization of East and West alike" (Atkins and Bodel 1972 p.27). Fever was recognized as a sign of illness several thousand years before the birth of Christ (figure 8.1), but the ancient Greeks also believed fever to be beneficial. This concept had its origin in the theory of Hippocrates that health and sickness were direct consequences of the interrelationships among the four body "humors"—blood, phlegm, yellow bile, and black bile. Galen is credited with the idea that fever derived from the accumulation of yellow bile, probably because of the frequency with which jaundice complicated the major diseases of his time: malaria, typhoid, hepatitis, and tuberculosis. Physicians of the time believed that fever "cooked" and separated, and ultimately evacuated, the excess humor by some combination of vomiting, diarrhea, and sweating (Yost 1950). This led physicians to "cure" their patients by administering drugs that increased body temperature or that served as emetics, purgatives, or sudorifics (Galen 1968).

The concept that fever is beneficial persisted for about two millennia. For example, in the seventeenth century, the English physician Thomas Sydenham wrote: "Fever is a mighty engine which Nature brings into the world for the conquest of her enemies" (Bennett and Nicastri 1960 p.16). However, it wasn't until the late 1800s, when C. Liebermeister correctly defined fever for the first time as the regulation of body temperature at a higher level; fever was considered to be dangerous because it was associated with reductions in body weight, appetite, and organ function.

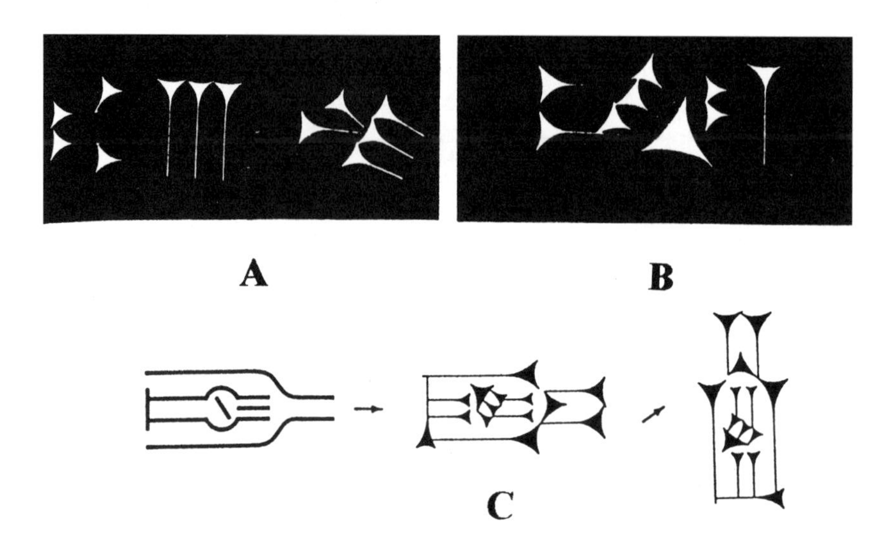

Figure 8.1
(A) Akkadian words meaning "fever" and "inflammation," *um* + *mu*. (B) *Ummu* written as a single ideogram. (C) A Sumerian word for inflammation. Here *ummu* appears inside a frame. When traced back to a Sumerian pictogram, the translation is "inflammation inside the chest." (Modified from Majno 1975.)

Interestingly, the idea that fever is harmful prevails today, despite the historical view and the preponderance of evidence to the contrary. As pointed out by Kluger (1979), this belief may be attributed to the fact that most antipyretic (fever-reducing) drugs also are analgesics (pain-reducing). Hence, feeling better after taking two aspirins may have little to do with reduction of body temperature, but is simply the result of relief of other symptoms of discomfort.

Evolution of Fever

Because descriptions of fever appear throughout recorded history, two intriguing questions immediately emerge: How did fever evolve? Does fever have survival value?

As Kluger (1979) said, we can not address the question of how fever evolved on the basis of studies of present-day animals. However, the observation that primitive animals alive today can develop fevers when infected with appropriate microorganisms strongly suggests that the

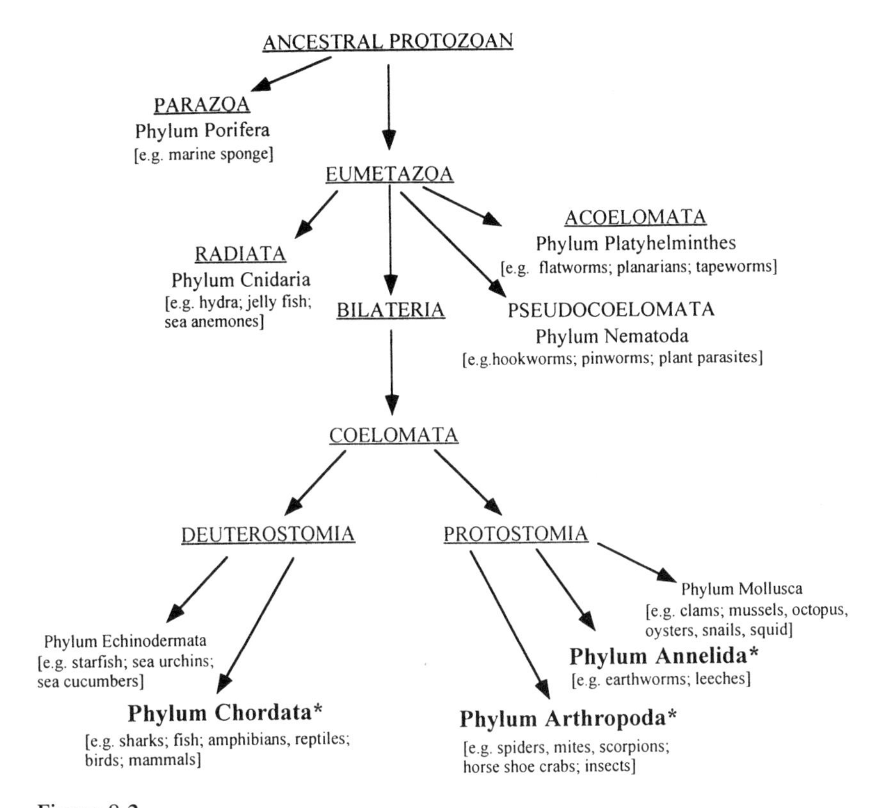

Figure 8.2
How fever evolved. Starred (*) phyla contain species that develop fever. (Modified from Kluger 1991.)

ancestors of these animals were capable of producing fevers. Based on the observation that representative animals from the phyla Chordata, Arthropoda, and Annelida are capable of developing fever, it is highly likely that this pathological response existed in animals of the two major divisions of the animal kingdom (Deuterostomia and Protostomia), and even in animals that predated them (figure 8.2). Ectothermic and endothermic invertebrates and vertebrates, with few exceptions, develop fevers when administered endotoxin or some other pyrogenic substance. Evidence in support of this statement is provided in table 8.1 (Kluger 1991). Although some reports indicate that several species do not develop fevers, these negative results should be viewed with caution because the animals in these instances may not have been infected with an appropri-

Table 8.1
Febrile responses of ectothermic vertebrates and invertebrates

Species	Activator of Fever
Reptiles	
Dipsosaurus dorsalis	Bacteria, endogenous pyrogen
Iguana iguana	Bacteria
Crotaphytus collaris	Bacteria
Terrepene carolina	Bacteria
Chrysemys picta	Bacteria
Sauromalus obesus	Bacteria
Alligator mississippiensis	Bacteria
Amphibians	
Hyla cinerea	Bacteria
Rana pipiens	Bacteria
Rana catesbeiana	Bacteria
Rana esculenta	Bacteria, prostaglandin PG E_1, endogenous pyrogen
Necturus maculosus	PGE_1
Fishes	
Micropterus salmoides	Bacteria
Lepomis macrochirus	Endotoxin, bacteria
Carassius auratus	Endotoxin, bacteria
Invertebrates	
Cambarus bartoni (crayfish)	
Gromphadorhina portentosa (cockroach)	Endotoxin, bacteria
Gryllus bimaculatus (cricket)	*Rickettesiella grylli*
Melanoplus sanguinipes (grasshopper)	Nosema acridophagus
Homarus americanus (lobster)	PGE_1
Penaeus duorarum (shrimp)	PGE_1
Limulus polyphemus (horseshoe crab)	PGE_1
Buthus occitanus (scorpion)	PGE_1
Androctonus australis (scorpion)	PGE_1
Onymacros plana (tenebrionid beetle)	Endotoxin
Nephelopsis obscura (leech)	Endotoxin, PGE_1

Modified from Kluger 1991b.

ate pyrogen for that species, or endogenous antipyretic agents (such as glucocorticoids) may have suppressed the release of endogenous pyrogen when the animal was stressed.

As noted by scholars who study evolution, fever is a metabolically expensive process. Heat production increases about 20 percent during fever primarily because the elevated temperature increases the rates of all biochemical reactions (the Q_{10} effect discussed in chapter 4). Thus, fever is a costly event, especially when considering that primitive animals constantly had to avoid predators and seek food to maintain metabolism. If fever did not have some adaptive value, it is probable that such an "expensive" process would have been eliminated ages ago. The fact that fever has such a long evolutionary history strongly suggests that it was important in ameliorating disease. What is the evidence that fever has survival value?

The Astonishing Experiments

If fever evolved over hundreds of millions of years, does that mean it necessarily benefits the host? If pathogen and host coevolved, one could argue that fever benefits the pathogen, not the host. How could we tell if fever contributed to survival or death?

Aside from mammals and birds, all other vertebrates and invertebrates have low rates of intrinsic heat production and therefore regulate their body temperature behaviorally. Although behavioral regulation is somewhat crude compared with the thermal homeostasis of mammals and birds, ectothermic species manage to control body temperature within relatively narrow limits. They achieve this by expressing a preference for a particular range of environmental temperatures. To increase or decrease their body temperature, they move to a warm or a cool environment, respectively. Because body temperature equilibrates with ambient temperature, the thermal preference of an ectotherm reflects its preferred body temperature.

Astonishingly, when ectothermic animals, such as the lizard, were infected with a bacterium and allowed to select an ambient temperature along a thermal gradient, which determines their body temperature, they selected a hot environment (50°C) that resulted in an elevation in their

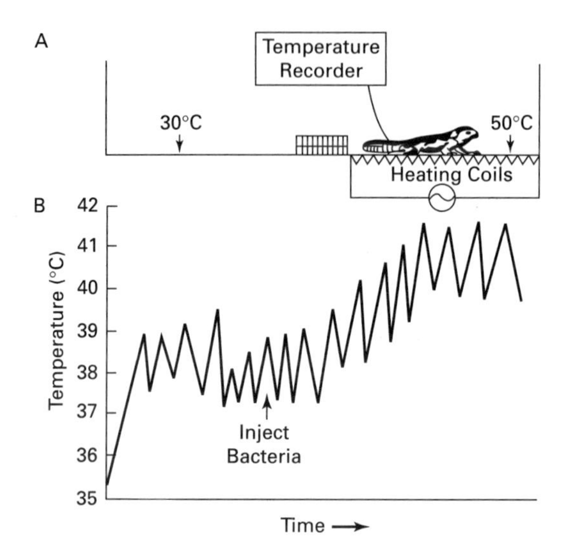

Figure 8.3
(A) Lizard (*Dipsosaurus dorsalis)* in a shuttle box maintained at 30°C at one end and 50°C at the other, with a bridge in the center. A thermocouple was inserted into the cloaca of the lizard to monitor body temperature continuously. (B)A continuous recording of the lizard's body temperature over time. Under control conditions, body temperature cycled between 39° and 37.5°C as the animal moved from the warm to the cool side of the box. This resulted in an average body temperature of about 38°C. When the animal was inoculated with pathogenic bacteria, body temperature rose to 41°C as a result of moving to the heated side of the box. Again the animal showed cycling behavior, but body temperature was now maintained at about 41°C. (From Kluger 1979.)

body temperature to 41°C rather than their normal preferred temperature of 38°C (figure 8.3). In essence, they behaviorally developed a 3°C fever. Even more astonishing is that if, after being infected, they were maintained at an ambient temperature of 38°C, 50 percent of the animals died, whereas only 15 percent died if they were allowed to migrate to an environment that elevated their body temperature to 40°C, and none died if they were allowed to migrate to an environment that elevated their body temperature to 42°C (Kluger 1979). These remarkable experiments dramatically illustrate the survival value of fever (Kluger 1991a).

Similar behavioral fevers have been observed in teleost fish (Reynolds and Casterlin 1976). When bluegill sunfish and largemouth bass were infected with *Aeromonas hydrophila* (gram-negative bacterium pathogenic

Table 8.2
Mean preferred temperatures (± 1 s.e.) before and after injection of *A. hydrophila* in ten fish

Species	s.l. (mm)	°C before injection	°C after injection	+ΔT (°C)
Micropterus salmoides	110	29.6 ± 0.3	31.9 ± 0.2	2.3
Micropterus salmoides	160	30.6 ± 0.3	32.3 ± 0.3	1.8
Grand mean (bass)		30.1 ± 0.5	32.2 ± 0.3	2.1
Lepomis macrochirus	120	31.9 ± 0.2	33.4 ± 0.2	1.5
Lepomis macrochirus	125	30.9 ± 0.2	32.4 ± 0.3	1.5
Lepomis macrochirus	130	31.6 ± 0.3	34.4 ± 0.3	2.8
Lepomis macrochirus	130	32.5 ± 0.2	37.2 ± 0.2	4.7
Lepomis macrochirus	145	27.9 ± 0.3	30.4 ± 0.2	2.5
Lepomis macrochirus	150	29.3 ± 0.3	32.2 ± 0.2	2.9
Lepomis macrochirus	150	30.8 ± 0.2	33.8 ± 0.3	3.0
Lepomis macrochirus	155	29.2 ± 0.2	31.4 ± 0.2	2.2
Grand mean (bluegill)		30.5 ± 0.6	33.2 ± 0.7	2.7
Grand mean (all fish)		30.4 ± 0.5	33.0 ± 0.6	2.6

s.l., standard length
From Reynolds and Casterlin 1976.

to mammals, reptiles, amphibians, and fish that causes hemorrhagic septicemia), they showed a mean increase of 2.6°C in mean preferred water (and therefore body) temperature (table 8.2). Control fish injected with a placebo showed no increase in preferred water temperature. Thus, lizards and fish exhibit a behavioral fever in response to a bacterial pyrogen that probably is the result of the pyrogen exerting its influence on the hypothalamus. Evidence to support the latter statement is that prostaglandin E (PGE) injected into the brains of various ectotherms produces a dose-dependent fever. And the astonishment continues. If an antipyretic such as acetaminophen is dissolved in the thermally graded water of sunfish, animals previously injected with bacteria remain normothermic (Moltz 1993).

The survival value of fever also has been demonstrated in mammals, including humans. In rabbits and mice infected with a pyrogen, increased survival occurred with fevers of about 2°C; higher fevers were associated with greater mortality. In patients who developed moderate (37.8–38.3°C) fevers from bacterial peritonitis and bacteremia, survival was greater than in patients who did not develop fever (Moltz 1993). Julius Wagner-Jauregg received the 1927 Nobel Prize in Physiology for his use of fever to counteract general paresis (syphilitic disease of the brain marked by progressive dementia, tremor, speech disturbances, and increasing muscle weakness). He developed the first successful therapy for syphilitic paresis of the insane by infecting patients with malaria. Psychiatric clinics and asylums throughout the world adopted the procedure, and the record of success was as high as 50 percent.

If fever enhances survival, what happens if you suppress fever? Is survival reduced? The answer is yes! Recall the experiment discussed above. When lizards were prevented from developing fever, more of them died. Moreover, if ectotherms and mammals (rabbits) are given aspirin before being infected with a pyrogen, fewer animals survive, and those that do survive, experience retarded recovery (Moltz 1993). Thus, the preponderance of data indicates that in both ectotherms and endotherms, fever has survival value.

What causes fever? For many years it was thought to be a failure of the thermoregulatory system. Does the brain actually regulate this response? Is there a ceiling for how high it can go? What are the circuitry and neurochemical basis for the generation of fever?

What Causes Fever?

It is easy to understand how physical exercise, which produces metabolic heat, can raise body temperature to 38°C–39°C; but how can a resting individual elevate body temperature to 40°C, or higher, in the generation of fever? Today, three different types of fever have been distinguished: pathogenic fever, the most common, which is initiated by microorganisms; neurogenic fever, produced by damage to brain tissue; and psychogenic fever, produced by "emotion-evoking stimuli" such as restraint or exposure to a unique environment (Moltz 1993). We will focus on pathogenic fever, the most commonly experienced. The important question here is How can a substance that causes fever produce the same elevation in body temperature, under resting conditions, as the elevation in body temperature that occurs with violent exercise?

Pathogenic fever is caused by microorganisms such as bacteria and viruses. Because these organisms cause fever, they are called pyrogens. If these pyrogens originate from outside the body, they are called exogenous pyrogens. When they are derived from the host, they are called endogenous pyrogens. The most potent endogenous pyrogen is endotoxin, which derives from the cell walls of gram-negative bacteria. (Bacteria are classified as gram-negative or gram-positive depending upon their ability to retain a blue dye, a test designed by the Danish physician Hans Christian Joachim Gram.) Endotoxins can be both harmful and beneficial to humans (figure 8.4). How best to harness the good and eliminate the bad is a current focus of extensive research. The cell walls of these bacteria contain lipid (fat), polysaccharide (carbohydrate), and protein. The toxic components are the lipid and polysaccharide. Hence, it is the lipopolysaccharide (LPS) component that activates leukocytes (monocytes and macrophages) to release a mixture of immunoregulatory polypeptides called cytokines. This mixture, which is considered to be endogenous pyrogen (EP), includes interleukin–1 (IL-1), interleukin–6 (IL-6), tumor necrosis factor (TNF), and interferon (IFN). *A fundamental concept in the pathogenesis of fever is that exogenous pyrogens cause fever only through their production of endogenous pyrogen.*

The first cytokine to be identified as an endogenous pyrogen was IL-1. Importantly, cytokines are not stored in macrophages; they are formed after macrophage stimulation. Furthermore, white cells of the mononuclear phagocyte lineage are not the only source of cytokines. Cytokines also are produced, in smaller quantities, by Kupffer cells of the liver, keratinocytes, fibroblasts, T cells, Langerhans cells, astrocytes, and endothelial cells. Therefore, the designations IL-1 and IL-6, which implies that these substances are derived solely from leukocytes, is misleading. IL-1 promotes leukocytosis, activates lymphocytes, stimulates prostaglandin synthesis, and has been implicated in the release of glucocorticoids, insulin, growth hormone, and thyroxin. Thus, it produces a variety of metabolic and hematologic effects that are collectively referred to as the "acute phase response." Fever occurs in many different patterns. The pattern observed will depend upon the substance or substances (enumerated above) that are synthesized and released, and where they act. Thus, it is appropriate at this stage to ask, What brain structures participate in fever production?

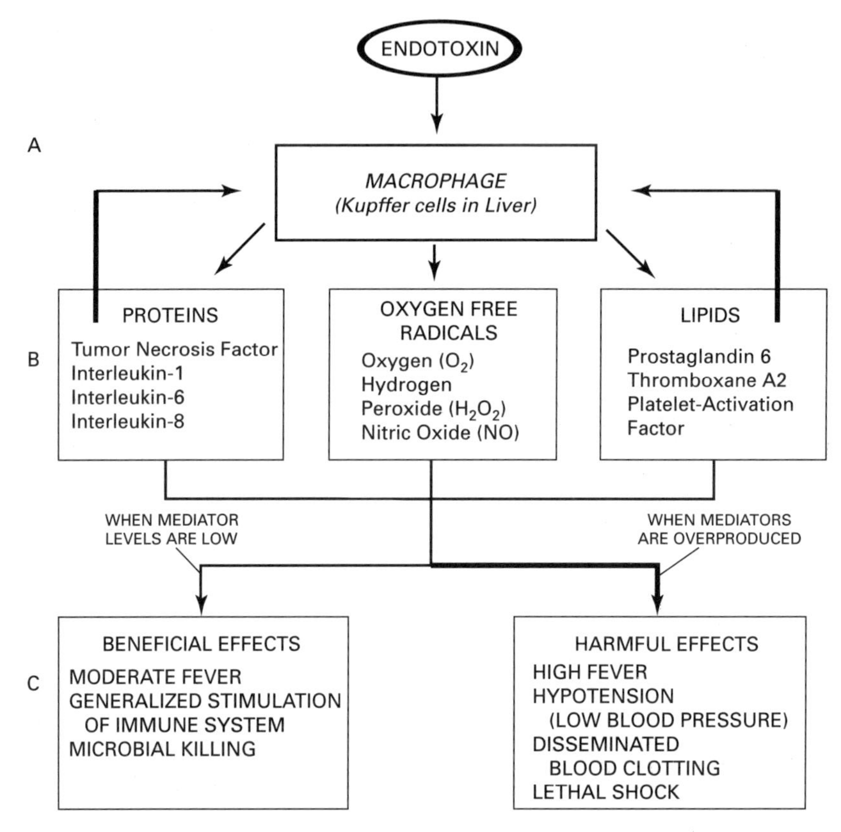

Figure 8.4
Endotoxin stimulation of macrophages (A) produces a variety of effects (B) mediated by the synthesis of proteins, oxygen free radicals, and lipids. These mediators may act in concert, independently, or in sequence to produce beneficial or harmful results (C). Mediator synthesis is enhanced by tumor necrosis factor and inhibited by prostaglandin E_2. (Modified from Rietschel and Brade 1992.)

Role of the Brain

It is now well documented, by placing very discrete lesions in different areas of the brain, that body temperature is controlled by a host of brain structures, including the hypothalamus, other structures of the limbic system, the substantia nigra, and the recticular formation. Because such lesions usually do not markedly impair thermoregulation, especially in animals not stressed by heat or cold, these experiments further indicate a certain degree of redundancy in thermoregulatory control and the presence of a hierarchy among neuroregulatory structures in this system. The brain site most often identified with thermoregulation is the preoptic anterior hypothalamus (POAH), because heat loss and heat gain mechanisms are so readily elicited when this area of the brain is heated or cooled, respectively. However, even lesions in this area of the brain do not completely impair thermoregulatory ability. The early concept of heat loss and heat gain centers in the brain is no longer tenable; nor is it clear if behavioral responses to thermal challenges (selecting a warmer or cooler environment, or modifying the clothing we wear)are controlled by areas of the brain different from those that control autonomic responses (sweating, shivering, vasomotor changes). Knowing the complexity of this system, which perhaps is the best example of homeostatic regulation in the human body, do we have any idea of where EP acts in the brain to produce fever?

As you might expect from the discussion above, lesioning the POAH does not impair one's ability to develop a fever. In humans, an adult with severe hypothalamic damage and an infant with the complete absence of the medial preoptic area (MPOA) were able to develop fevers following an infection. Monkeys, goats, and rabbits also exhibited fevers following the destruction of the MPOA. Moreover, rats that had their MPOA separated from the rest of the brain by bilateral knife cuts produced fevers in response to bacterial infection that were indistinguishable in magnitude and duration from those in sham-operated control animals. Thus, it appears that the MPOA is not essential for the generation of fever.

Injecting EP into the midbrain reticular formation, medulla oblongata, lateral hypothalamus, or pons will generate a fever, but of lesser magni-

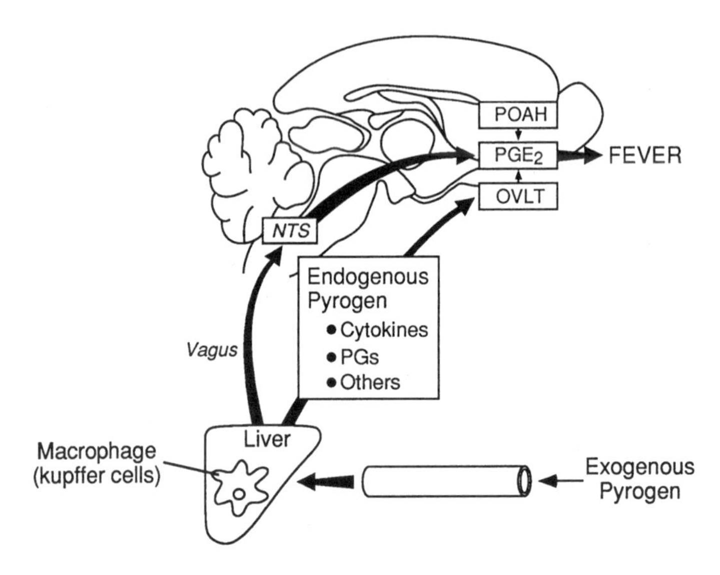

Figure 8.5
Schematic diagram illustrating the proposed sequence of events that produce fever. Exogenous pyrogens gain access to the circulatory system and cause the release of cytokines from Kupffer cells in the liver. These cytokines activate vagal afferents that terminate on noradrenergic cell groups in the nucleus tractus solitarius (NTS). The signal is then communicated to the preoptic anterior hypothalamus/organum vasculosum laminae terminalis (POAH/OVLT), where the release of norepinephrine (NE) stimulates the synthesis of prostaglandin PG E_2, which causes body temperature to rise. (Modified from Blatteis and Sehic 1997.)

tude as the distance from the POAH increases (Blatteis 1984). Interestingly, behavioral responses are elicited when EP is injected into the POAH and lateral hypothalamus, but not when it is injected into the pons or medulla oblongata, thus suggesting discrete brain loci for behavioral and autonomic regulation of fever. However, experiments using microknife cuts separating different areas of the brain indicated that such a distinction was too simplistic.

Do fever-generating cytokines appear in the brain? Do they cross the blood-brain barrier (BBB) and affect brain cells? What happens if they are injected into the brain directly? Do they produce the same pattern of fever as that produced by the more normal route of fever production; that is, when these substances are released into the blood? Although there is evidence that some cytokines can cross the BBB, the amount that crosses in

10 min appears to be inadequate to account for the fevers generated by these substances when they are introduced intravenously. Similarly, the delayed onset latency and slower rise in body temperature produced by introducing these cytokines into the cerebral ventricles rather than into the blood indicates that they do not enter the brain through leakage or facilitated transport at the choroid plexuses. Thus, they do not seem to exert their influence by first gaining access to cerebrospinal fluid. Perhaps the most popular hypothesis for the generation of fever involves the organum vasculosum laminae terminalis (OVLT) hypothesis. The OVLT is a circumventricular (in the area of a cerebral ventricle) organ that lacks a BBB and is located in the anteroventral wall of the third ventricle (AV3V) on the midline of the MPOA. The hypothesis states that blood-borne cytokines enter the OVLT and diffuse to the POAH, where they release prostaglandin E_2 (PGE_2) to cause fever. The observation that lesioning the AV3V region prevents the febrile response to EP, without impairing normal temperature regulation, provides strong support for this hypothesis. What remains unclear is the precise origin and site of action of PGE_2. In addition to this more traditional means of producing fever, recent studies indicate that activation of subdiaphragmatic vagal afferents also can produce fever (Blatteis and Sehic 1997).

The most current theory of how a rapid fever is produced by blood-borne LPS is illustrated in figure 8.5: LPS in the blood activates certain serum proteins, called complement, that bind to hepatic macrophages (Kupffer cells); Kupffer cells release one or more cytokines and/or PGE_2, which are capable of activating adjacent subdiaphragmatic vagal afferents; these afferent signals are transmitted to noradrenergic (NE) cell groups within the nucleus tractus solitarius (NTS); and these NE signals are transmitted to the POAH/OVLT, where NE is released and, in turn, releases PGE_2. The PGE_2 then acts to increase the set-point, which signals heat production and heat conservation mechanisms (Blatteis and Sehic 1997).

How does the brain bring about these increases in heat production and heat conservation? It is apparent that several areas of the brain are involved in the generation of a fever because when the OVLT, POAH, or MPOA is destroyed, intravenous endotoxin, can still produce fever. The postulated mechanism is that endotoxin, via PGE or some other interme-

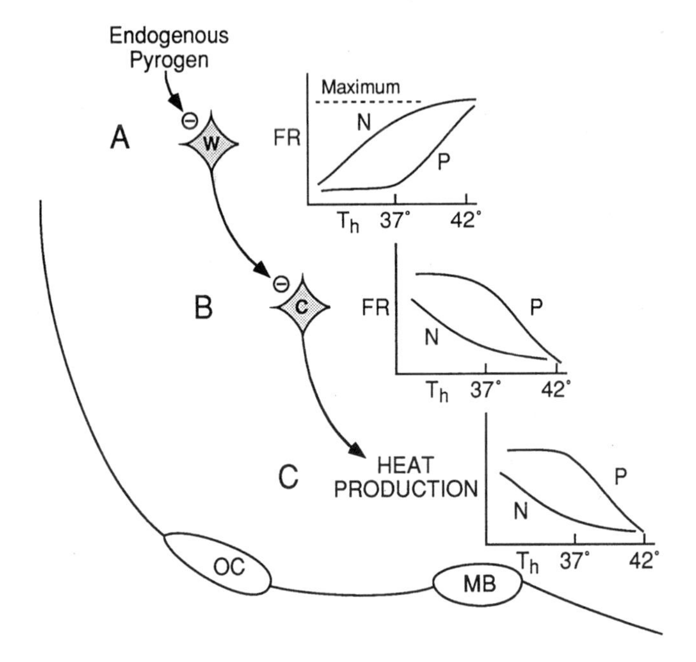

Figure 8.6
The effects of endogenous pyrogen (P) on the firing rate (FR) of warm (W)-sensitive (A) and cold (C)-sensitive (B) neurons and whole-body heat production (C). Under normal (N) conditions, increasing hypothalamic temperature (T_h) increases the FR of warm-sensitive neurons; some of these synaptically inhibit (-) other neurons, making them respond as if they were cold-sensitive (negative slope). By inhibiting (-) the FR of warm-sensitive neurons, endogenous pyrogens increase the FRs of cold-sensitive neurons and heat production. (Modified from Mackowiak and Boulant 1996.)

diate, inactivates warm receptors while activating cold receptors in the POAH. These neurons subsequently stimulate neuronal pools responsible for activating the shivering pathway and increased constriction of cutaneous blood vessels. Behavioral responses, such as moving to a warmer environment and donning extra clothing, also occur. Figure 8.6 illustrates a neuronal scheme to produce fever by increasing heat production. Warm- and cold-sensitive neurons receive a host of afferent inputs from skin, spinal cord, and other central sources. However, a fascinating aspect of warm-sensitive neurons is that they retain their thermosensitivity even after their synaptic input is blocked, that is, they are intrinsically thermosensitive. In contrast, cold-sensitive neurons usually lose their cold sensitivity following synaptic blockage. Under normal conditions, warm-sensitive neurons show near maximal firing rates at 37°C, which produces near maximal inhibition on neighboring cold-sensitive neurons, and heat production is markedly suppressed.

Electrophysiological studies indicate that pyrogens inhibit warm-sensitive neurons, excite cold-sensitive neurons, and have little or mixed effects on temperature-insensitive neurons. Note that in figure 8.6A, at 37°C, pyrogen has a maximal inhibitory effect on warm-sensitive neurons. Because warm-sensitive neurons inhibit cold-sensitive neurons, maximal inhibition of warm-sensitive neurons by pyrogens at 37°C produces maximal stimulation of cold-sensitive neurons at 37°C. This results in maximal heat production at 37°C and coincides with the chill phase of fever. As core temperature rises to reach a new elevated set-point, heat production diminishes. At 42°C, the firing rate of warm-sensitive neurons reaches its zenith, whereas cold-sensitive neurons reach their nadir, suggesting that regulated increases in core temperature above 42°C may not be possible because thermosensitive neurons can no longer provide input to regulate effector responses.

Is the MPOA the brain area most sensitive to the presumed mediator of fever, PGE? The evidence shows that PGE injected into other areas of the brain (ventromedial hypothalamus, for example) generates a more robust fever than is demonstrated by PGE injected into the MPOA, and that the density of PGE binding sites in the MPOA is low. Thus, the MPOA may not be the major brain site required to produce fever. PGE can activate multiple hypothalamic nuclei as well as extrahypothalamic loci, including the corpus striatum and ventral septal area (Moltz 1993). It may be that a neuronal network is required to produce the autonomic and behavioral responses characteristic of fever.

Fever's "Glass Ceiling"

The term "glass ceiling" is taken from an intriguing paper by Mackowiak and Boulant (1996). The question is "Does fever have a relatively "fixed" ceiling that can not be exceeded, or can this presumed upper limit be penetrated, resulting in excessively high temperatures that can be fatal? The upper limit for a "regulated" fever is 41°C–42°C (DuBois 1949), but this upper limit varies among mammals, the body site where temperature is measured, and according to the infectious disease acquired. Malaria is notorious for producing high fevers, whereas HIV results in little if any fever. The need for an upper limit seems obvious—above the upper limit, the pathogenic microorganisms die, but so may the host. Kluger found that when lizards were infected with a natural pathogen that afflicts them, there was a correlation between temperature and survival rate when body temperature was increased from 34°C to 40°C. However, when body temperature rose to 42°C, uninfected control animals died.

Aspirin: Are We Using It Correctly?

Imagine having hot and cold chills, a runny nose, a headache, and an oral temperature of 39°C. Self-diagnosis: a common cold with fever. Self-prescription: drink plenty of fluids, take two aspirins, and go to bed. You proceed to the medicine cabinet, remove the bottle of aspirin, and pop two tablets in your mouth. Have you just made a mistake? Aspirin suppresses endotoxin- and cytokine-induced fevers not only by inhibiting cyclooxygenase, the enzyme responsible for synthesizing PGE, but also by promoting the release of AVP (Alexander et al. 1989).

The harmful and/or beneficial effects of the febrile response can be attributed to (a) the rise in core temperature and/or (b) the effects of the febrile mediators (IL-1, IL-2, IL-6, interferon, and TNF). We have already alluded to the survival benefits of the rise in core temperature associated with fever in reptiles. Retrospective analysis of clinical data shows improved survival in patients who developed relatively high fevers in response to bacteremia, polymicrobial sepsis, and spontaneous bacterial peritonitis (Mackowiak 1994). In addition to the benefits associated with elevations in core temperature, other investigators have shown enhanced resistance to infection produced by the effects of the endogenous mediators (IL-1, IL-2, IL-6, TNF, and interferon) of the febrile response (Mackowiak 1994). These beneficial effects of fever and its mediators are limited and are directed primarily against localized infection. This evidence combined with the observation that fever occurs in vertebrates from fishes through mammals, is compelling evidence that fever is beneficial and that the use of antipyretics (e.g., aspirin) may not be the most effective treatment for diseases that produce fever. In some instances, fever perhaps should be allowed to run its natural course, or even be elevated.

The febrile response also has been viewed as harmful. For example, administration of an anti-interferon antibody significantly reduces mortality in response to the induction of septic shock; and bacterial sepsis can be attenuated by pretreating animals with IL-1 antagonists and monoclonal antibodies directed against TNF. Thus, there is growing interest in developing agents to combat the effects of pyrogenic cytokines.

If fever is beneficial, what are the mechanisms responsible for its adaptive value? Figure 8.7 illustrates a number of possibilities. First, the beneficial effect may simply be that a high temperature kills the invading microorganisms. For example, the bacteria that cause gonorrhea, pneumococci bacteria, and spirochetes that cause neurosyphilis are destroyed by elevations in temperature to 41°C (Kluger 1979). In addition to this direct effect, there are several indirect effects of fever that can benefit the host. These effects include increased lysosomal function, inhibition of viral growth and synthesis by increased release of interferon, and increased leukocyte function. Leukocytes are white blood cells that play a crucial role in immune function. Neutrophils, which are a form of while blood cell, move through cell membranes more effectively, phagocytize, and kill bacteria more effectively at fever temperatures than at 37°C.

There also is some evidence that antibody production and lymphocyte transformation are enhanced by fever temperatures. Although there clearly is much to be learned about immune function and how it is affected by fever, the evidence supports a positive role for fever in the management of disease. Evidence also indicates a role for cytokines in host defense. Although they are not fever-dependent, febrile temperatures do enhance the effect of IL-1 and TNF or T lymphocyte proliferation, and the effect of IL-1 on reducing serum iron concentration. Because almost all bacterial pathogens require iron, hypoferremia is considered a host-defense mechanism and has been termed "nutritional immunity," to characterize the attempt of the host to deprive invading pathogens of needed iron.

In the final analysis, is fever a blessing or a curse? Mackowiak (1994) offers a unifying hypothesis that requires the febrile response to be viewed

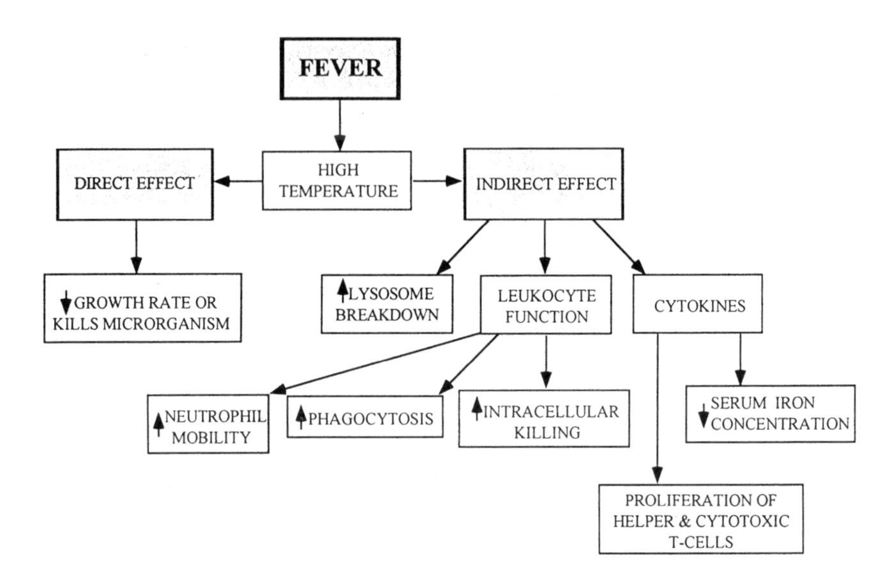

Figure 8.7
Potential mechanisms of host defense and survival. (Modified from Kluger 1979.)

as a mechanism to preserve the species rather than save the individual. Accordingly, fever and its mediators evolved to enhance the recovery of those afflicted with mild to moderately severe infection, in order to help propagate the species and to eliminate those with fulminating disease, who pose a threat of causing an epidemic that threatens survival of the species.

Does the Brain Produce Its Own Aspirin?

If we accept the concept that fever has an upper limit, then regulatory processes that prevent T_C from exceeding 41°C to 42°C must exist. Does the brain produce aspirin-like drugs (antipyretics or endogenous cryogens) that can antagonize the temperature-elevating effects of pyrogens? "Antipyretics" are defined as substances that reduce febrile temperatures but have no effect on normal core body temperatures. The story of antipyretics began with observations made on periparturient ewes and their newborn lambs. Five days before term, the pregnant ewes showed an attenuated response to endotoxin-induced fever, and within 24 hours of term, no fever was observed (Kasting et al. 1978). Moreover, the newborn lambs also were fever resistant, suggesting that mothers and lambs possessed some antipyretic substance near term. Arginine vasopressin (AVP), a peptide hormone synthesized in hypothalamic neurons and stored in the posterior pituitary gland, emerged as a candidate when elevated plasma concentrations of AVP correlated well with the diminished responsiveness of pregnant sheep and their offspring to fever-inducing agents. Subsequent studies showed that (a) microinjections of AVP into select brain sites attenuate pyrogen-induced fever, and (b) antagonists to AVP receptor block the antipyretic response (Kluger 1991b; Moltz 1993). Furthermore, infusing hypertonic saline or inducing hemorrhage—both potent physiological stimuli causing AVP release—causes antipyresis. On the other hand, castration, which reduces AVP concentrations in the brain, exacerbates a PGE-induced fever (Kluger 1991b; Moltz 1993).

Another potential antipyretic, at low doses, is alpha-melanocyte stimulating hormone (**α**-MSH). It is named for its ability to darken the skin, but it is over 25,000 times more potent than acetaminophen as an antipyretic when administered centrally (Mackowiak and Boulant 1996). Evidence that (**α**-MSH may function to "help set fever's upper limit" is that the highest concentrations of it in septal perfusates were found during the rising phase of fever (Bell and Lipton 1987). Another endogenous agent that may function as an antipyretic is corticotrophin releasing factor (CRF). The cytokine IL-1 stimulates the release of CRF from the hypothalamus, which in turn releases glucocorticoids from the adrenal cortex. The glucocorticoids inhibit pyrogenic cytokines including IL-1, IL-6, and TRF. Thus, CRF and glucocorticoids may participate in a negative feedback loop to help set fever's upper limit. How these different antipyretics function together to "control" fever is not understood, and should be the focus of future research.

Does Exercise Produce Fever?

Is the rise in body temperature during exercise a fever? Before we can address this question, it is critical to define "fever" and distinguish it from "hyperthermia."

Core body temperature rises during both fever and exercise, but as a fever develops, we shiver, whereas during exercise we sweat. How is this possible? In both situations we become hyperthermic (body temperature rises). How can the same elevation in body temperature produce opposite responses? To understand this phenomenon, body temperature can be placed in one of four categories depending upon the concept of set-point (Snell and Atkins 1968), which was described in detail in chapter 2. These categories are (1) normothermia—measured body temperature and the set-point coincide; (2) hypothermia—measured body temperature is below the set-point, which may or may not be normal; (3) hyperthermia—measured body temperature is above the set-point, which may or may not be normal; (4) fever—the set-point is raised, but measured body temperature may or may not be raised to the same level. Set-point is defined as that value of the controlled variable at which the control action is zero (Hensel 1981). Thus, core temperature (the controlled variable) is at its set-point when the organism is neither heating nor cooling (controlled action) itself. During the development of a fever, heat production, heat conservation, and behavioral mechanisms are activated until core temperature is elevated to the newly established set-point (figure 8.8).

Once core temperature reaches the new set-point, it is regulated around this elevated level. The concept that fever is regulated is based on Liebermeister's observation that a febrile patient returns body temperature to a febrile level after experimental warming or cooling. When the fever breaks, the set-point returns to its normothermic level, the patient feels warm because body temperature is above the set-point, and heat-loss mechanisms are activated to return body temperature to its normothermic level. When the set-point increases as a result of endogenous pyrogen, blood is shifted away from the skin, thus adding to the skin's insulation, and heat production is increased by shivering, which increases the tone of skeletal muscles. Note the associated behavioral response: adding clothing and moving to a warmer ambient temperature. When muscle tone exceeds a certain threshold, the muscles tremble. Heat production during shivering can increase to four to five times resting values.

Fever can occur at any ambient temperature. Moreover, the rise in body temperature during a fever that occurs when ambient temperature is 50°F is virtually the same as the rise that occurs when ambient temperature is

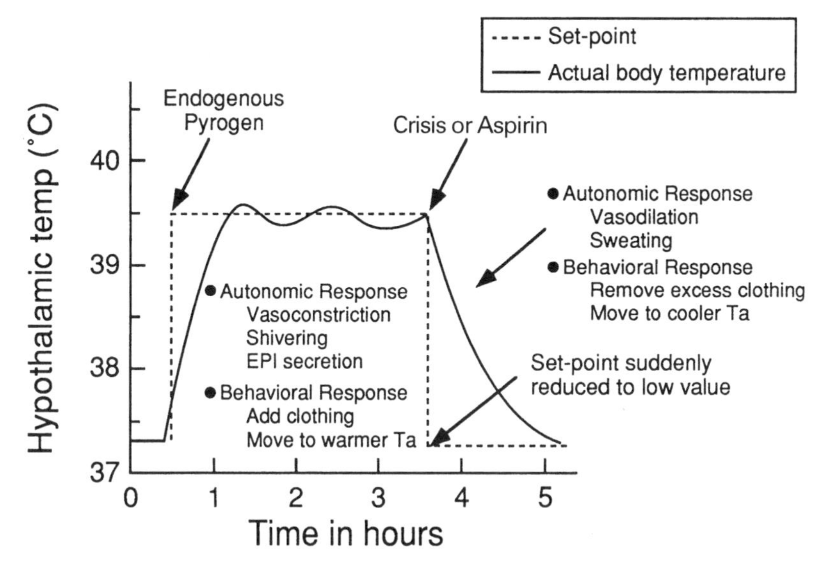

Figure 8.8
Effect of pyrogens on the set-point and the mechanisms responsible for the generation of fever. Note that following the rise in set-point, mechanisms to increase heat are activated (heat production, vasoconstriction, behavioral response), but that when the fever breaks (crisis) or following the ingestion of aspirin, the elevated temperature is perceived as hot and the patient activates heat loss mechanisms (sweating, vasodilation, behavioral response.)

75°F. How can this be? In the cold, mechanisms for conserving and producing heat increase, while in a warm environment, the mechanisms for losing heat decrease. In other words, in the cold, heat production increases and skin blood flow decreases, whereas in the heat, sweating and skin blood flow decrease. During the onset of a fever, a room temperature that was comfortably warm, feels cold. This signifies that the perception of temperature is altered so as to reinforce the drive to raise core temperature to a higher level.

Thus, unlike other forms of hyperthermia, the rise in body temperature during a fever is regulated at a new set-point. And, unlike other forms of hyperthermia, where whole-body cooling is the only effective treatment for lowering body temperature, cooling a febrile patient will be countered by physiological mechanisms to maintain the elevated set-point. Only aspirin-like drugs, which presumably block the effects of pyrogen in the brain, reduce the set-point, which subsequently allows the body to lower its temperature back to its normal control condition. Giving aspirin to an athlete has no effect on the rise in body temperature during vigorous exercise. The only method of lowering body temperature during the hyperthermia of exercise is by whole-body cooling.

During exercise, the rise in core body temperature often has been compared with turning up the thermostat in a house. Such a description would be analogous to the effect of pyrogen. Are pyrogens produced during exercise? Some recent data suggest that this may occur. Can exercise produce endotoxemia, and if so, what effect might this have on brain function, especially areas concerned with thermoregulation?

Figure 8.9 compares the hyperthermia caused by fever and exercise. During fever, the elevation in temperature is actively produced and regulated at a new set-point. During exercise, the increase in body temperature occurs passively, from the accumulation of heat in the working skeletal muscles; blood flow to the skin increases, and we begin sweating. During exercise the set-point does not change. Body temperature rises to a plateau when heat production equals heat loss, but this new elevated body temperature is not regulated.

If the same intensity of exercise (and therefore heat production) is performed in cooled stirred water and in air at the same temperature, the rise in body temperature in the water will be much less because water has a much greater capacity to remove heat from the body by conduction. This simple experiment illustrates that body temperature during exercise is not regulated, but is simply the result of the balance between heat production and heat loss. In fact, if exercise is performed in cold water and heat loss exceeds heat production, body temperature will fall during exercise.

If the rise in body temperature during exercise were the result of fever, taking aspirin should prevent this rise. Taking aspirin prior to or during exercise has no effect on the rise in body temperature in humans, which convincingly demonstrates that the set-point does not rise during exercise. Are there any circumstances under which exercise might release cytokines and produce fever? Yes, there are. During ultraendurance events such as the triathlon, there have been reports of exercise initiating an acute phase response, endotoxemia, and a decrease in anti-LPS immunoglobulin G (see Kluger 1991b: 115). The question of whether exer-

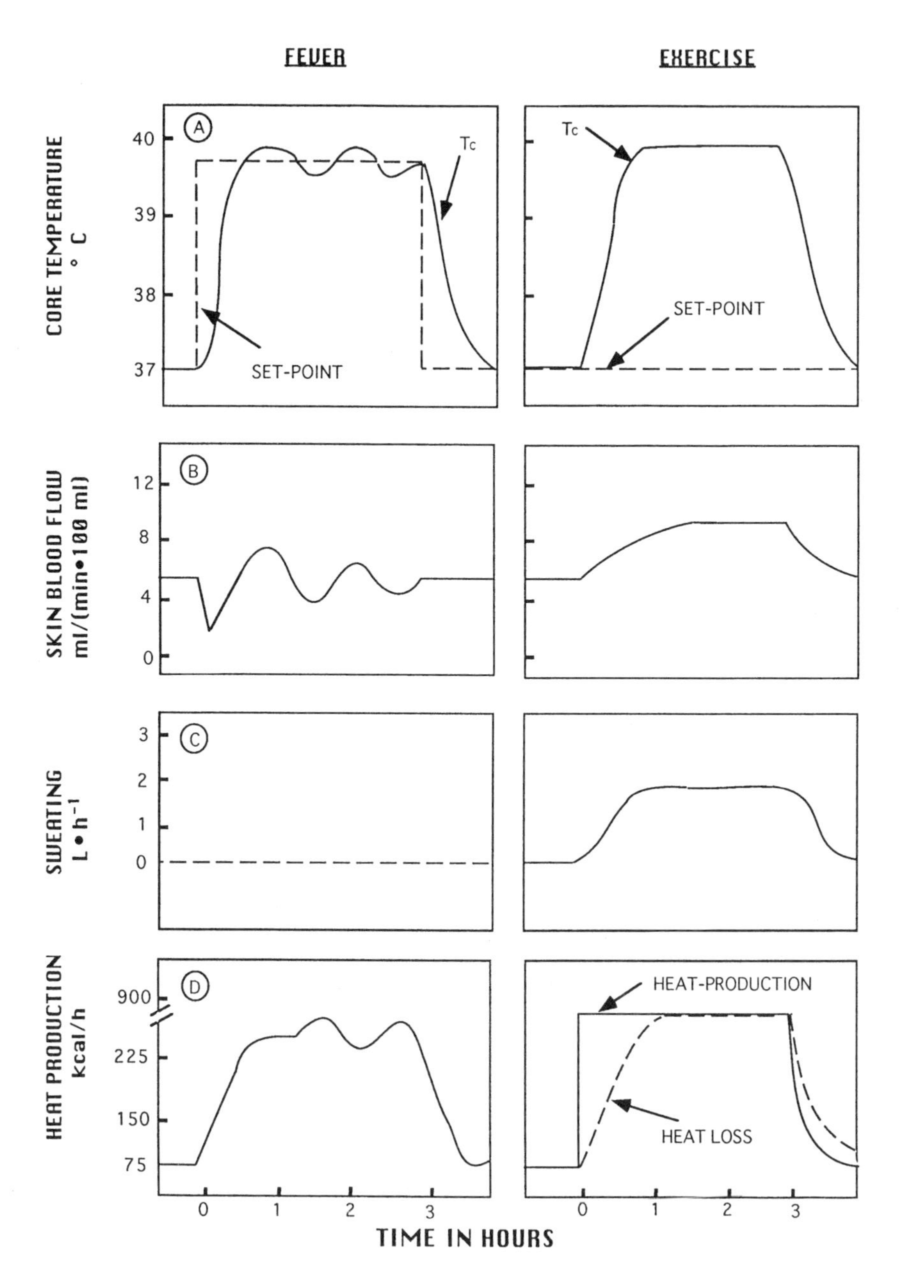

Figure 8.9
Fever increases the set-point, and body temperature must rise to reach a new thermal balance. During exercise, the set-point does not change. Body temperature is elevated by increased heat production until the rate of heat loss rises sufficiently to achieve a new thermal balance.

cise produces fever in experimental animals is not so clear. The desert iguana (an ectotherm), for example, which regulates body temperature behaviorally by selection of preferred environmental temperatures, chose a warm environmental temperature following treadmill exercise, thus suggesting an elevated set-point. Most interesting, the administration of an antipyretic to these animals eliminates the rise in postexercise body temperature.

Another interesting difference between fever and exercise is the ambient temperature preferred—also called "thermal preferendum"—during fever and exercise. If we had the choice of cool or warm air being blown across our skin as body temperature was rising during fever and during exercise, we would select warm air during the onset of fever, but cool air during exercise. Thus, in fever we would facilitate the rise in body temperature, but during exercise we would facilitate heat loss In summary, fever is actively produced, regulated, and reduced with aspirin. The rise in body temperature during exercise occurs passively, is not regulated, and is not affected by aspirin.

Age and Fever

Human neonates (first month of life) can only maintain a stable body temperature within only very narrow environmental conditions and, interestingly, usually do not develop fevers in response to infection. However, under severe conditions—very high doses of the infecting organisms—they become febrile (Moltz 1993), indicating they have an elevated fever threshold. How can we explain this phenomenon? Could it be that neonates produce less endogenous pyrogen in response to endotoxin? Or perhaps the immune system must first be sensitized to EP before fever can develop. In other words, an initial dose of EP may not produce fever, but a subsequent dose will. Because leukocytes from neonates incubated with endotoxin produce similar amounts of EP as adult leukocytes, and because some neonates develop fever when initially exposed to endotoxin, these possibilities seem unlikely.

Could the explanation be the brain? Is the thermoregulatory machinery responsible for producing fever insensitive to cytokines and/or PGE? Most of the experimental work on neonates has been conducted on lambs and guinea pigs. In lambs, infusing PGE into the brain failed to produce fever, but aspirin was effective in reducing fever produced by endotoxin, suggesting that PGE is involved in neonatal fever. Thus the evidence is confusing. It is not clear where in the brain PGE may be acting to produce fever.

Another intriguing possibility is that the neonate possesses high concentrations of its own endogenous antipyretic. As mentioned above, AVP is a potent antipyretic, and in the neonatal brain, the concentrations are exaggerated.

Fever in the Aged

An often quoted statement by Wunderlich (see Lipton and Ticknor 1979 p. 269) regarding fever in the aged is the following:

"In very old people, it is common in various diseases to find the temperature half a degree or more lower than is common in similar circumstances in younger people." However, when retrospective reports on humans were examined and the method, site, and frequency of temperature measurement were evaluated, a difference between old and young people was difficult to confirm. For example, core temperature in the elderly often is taken orally, which typically records 0.3° to 0.6°C lower than rectal temperature. Data collected from experimental animals also are inconsistent. A pattern that does emerge in animals is that the occurrence of fever in older animals shows a delayed rise. Despite the inconsistencies, the concept has emerged that the aged have lower fevers, and investigators have explored potential mechanisms to explain this phenomenon. Interestingly, older animals do not produce less cytokine in response to endotoxin challenge, nor do they possess less endogenous antipyretics (AVP, α-MSH); however, it has been observed that aged animals are more sensitive to α-MSH, to accommodate reduced concentrations of this putative neurotransmitter in the brain (see Moltz 1993).

An intriguing hypothesis regarding aging and homeothermy was proposed by Sohnle and Gambert (1982). Thcse investigators hypothesized that the ability to regulate body temperature evolved as an advantage against aging. Maintaining body temperature below optimal immunological function may serve as a mechanism to reduce the role of the immune system in the aging process.

9

Temperature and the Struggle for Life

In addition to heatstroke and fever producing high elevations in core body temperature (CBT), there are two other fatal syndromes that can raise CBT to levels even higher than those typically observed with heatstroke and fever. These are malignant hyperthermia (MH) and neuroleptic malignant syndrome (NMS). As the word "malignant" signifies, these syndromes originally were associated with high rates of morbidity and mortality. Today, successful treatment has significantly reduced the number of patients in whom the incidence of these disorders is fatal. Does either of them originate in the brain? Exertional heatstroke often is confused with both of them. Can exercise trigger one or the other? In recent years, a third syndrome, equally fatal and more rapid in onset than NMS, is serotonin syndrome (SS). SS and NMS present almost identically, but SS occurs within hours of receiving the precipitating agent, whereas NMS develops three to nine days after treatment is initiated. Is this latter syndrome the result of an imbalance in brain chemicals or substances produced elsewhere in the body? If hyperthermia is associated with such high rates of morbidity and mortality, how is it that CBT can be elevated to 41°C or higher in the treatment of cancer without producing brain lesions and/or the denaturation of protein?

The Engine Burns Out: Malignant Hyperthermia

In 1960, an Australian physician named Denborough was interviewing a 21-year-old student who was about to have surgery for a compound fracture of his leg. During the discussion, the young man expressed concern that 10 out of 24 of his relatives had died during a surgical procedure requiring general anesthesia. Given this information, Denborough chose halothane rather than ether as the general anesthetic. Within 10 min of the administration of halothane, the young man experienced an increase in heart rate and a drop in blood pressure, and his CBT began to rise sharply. The anesthetic was immediately stopped, and the patient was packed in ice. When the hospital records of his relatives who died during surgery were examined, they all showed a marked elevation in body temperature.

At approximately the same time that Denborough made his observations, a similar syndrome was observed in pigs that were bred for rapid growth and lean muscle. The syndrome in these animals was triggered by stimuli, such as shipping in confined quarters, that produced fighting among the animals, and was characterized by accelerated metabolism, acidosis, muscle rigidity, high temperature, and death. The benefit derived from this unfortunate syndrome in pigs was the development of an animal model to study the disease and the early development of a treatment.

Malignant hyperthermia is a rare genetic myopathy that affects about 1 in 15,000 children and 1 in 40,000 middle-aged adults (Johnson and Edleman 1992). It is triggered primarily by anesthetic agents (succinylcholine, halothane, isoflurane, enflurane), but on rare occasions it has been triggered by workplace chemicals, street drugs, X-ray contrast dyes, violent exercise, muscle trauma, fever, infection, pain, shivering, and agitation (Britt 1996). It is a defect in the ryanodine (RYR1) gene. When the syndrome is triggered, there is a prolonged opening of the ryanodine pores, resulting in a marked release of Ca^{2+} from the sarcoplasmic reticulum into the cytoplasm through defective ryanodine receptors. Muscle rigidity occurs in the jaw, then spreads to other skeletal muscles. This is a hypermetabolic syndrome that produces massive quantities of heat. CBT can rise 1.8°F (1°C) every 5 minutes.

Once diagnosed, MH patients can be safely anesthetized with nitrous oxide, barbiturates, and other anesthetics. Dantrolene, a drug that decreases calcium release from the sarcoplasmic reticulum, also can be used as a preventive measure. Whereas the mortality among MH patients in 1960 was 70 percent–90 percent, correctly diagnosed and treated persons today have almost a 100 percent chance of recovery from an MH crisis. All operating room personnel should have fundamental knowledge of this syndrome. Because it also can occur postoperatively, medical-surgical nurses should be skilled in identifying its pathophysiology, signs, and symptoms (Young and Kindred 1993).

Neuroleptic Malignant Syndrome

A 60-year-old man, treated with neuroleptics, was transferred from the psychiatric ward to the medical service of City Hospital on August l, 1997, because of hyperthermia (core temperature of 40.4°C). In the medical service no infection was found and the patient was treated with haloperidol. This is a typical case in which neuroleptic treatment and hyperthermia are correlated.

The pathogenesis of NMS is unknown. It has been described since 1960, primarily in the French psychiatric literature. In the United States, a few cases were described in the 1950s and 1960s, but most cases were misdiagnosed. NMS is clinically similar to MH, but is pharmacologically distinct. It occurs primarily in patients with psychiatric illnesses, but the full-blown syndrome also has been observed in normal persons treated with these neuroleptic drugs as preinduction anesthetics. It is characterized by four clinical hallmarks: (a) hyperthermia, (b) muscle rigidity, (c) mental status changes, and (d) autonomic instability. Muscle hypertonicity, described most often as generalized "lead pipe" or "plastic" rigidity, develops concomitantly with akinesia shortly before temperature elevations as high as 42°C. Consciousness may fluctuate from alert but dazed mutism to stupor and coma. Autonomic nervous system involvement is manifested by tachycardia, labile blood pressure, and incontinence.

The signs of NMS can occur from hours to months after the initial drug exposure; once initiated, they develop explosively in 24 to 72 hours. Rigidity, a cardinal feature of NMS, clinically distinguishes it from heatstroke. It has been associated with drugs that exert a dopaminergic blocking effect on the basal ganglia and hypothalamus, which has led to the hypothesis that it results from an alteration of central neuroregulatory mechanisms and to therapies directed at receptor sites in the brain. Dopamine is a neurotransmitter that activates heat loss mechanisms. Administration of a dopamine blocker will have the effect of elevating body temperature. (See Chapter 3 for additional information on this and other neurotransmitters.)

The Hyperthermia of "Ecstasy": Serotonin Syndrome

A 19-year-old student collapsed with a grand mal seizure after taking 3,4-methylene dioxymethamphetamine (MDMA), known as "Ecstasy," an increasingly frequent drug of abuse. Upon admission to the hospital she was unresponsive with a peripheral temperature of 40°C, displayed rigidity, and had an elevated creatine kinase concentration in the blood. A CT scan of the brain showed cerebral edema and toxicological studies of the blood confirmed the presence of large amounts of "Ecstasy." In contrast to NMS, autonomic instability was not observed. After treatment with diazepam (Valium) and dantrolene, her temperature returned to normal within 10 hours and she recovered. This fortunate young woman experienced serotonin syndrome, a recently discovered hyperthermic disorder very similar to NMS.

Serotonin syndrome is a hyperthermic reaction to drugs that increases serotonin levels in the central nervous system. This can be accomplished by increased release of serotonin, which is the effect of MDMA, or by blocking the reuptake of serotonin once it has been released, which is the effect of fluoxetine (Prozac). Other psychopharmacologic agents that enhance serotonin neurotransmission include monoamine oxidase inhibitors (MAOI), which block the breakdown of serotonin causing a presynaptic build-up of serotonin.

Stress-Induced Hyperthermia

The following example illustrates stress-induced hyperthermia. Suppose there is a group of 10 male mice housed together in a cage. Each minute a mouse is removed from the cage to measure its rectal temperature, but it is not returned to the group. With each mouse that is removed from the group, the temperature of the next mouse selected increases up, to the 8th mouse. This reliable and stable phenomenon, called stress-induced hyperthermia, produces a maximal rise in temperature of 1.3°C–1.8°C (Groenink et al. 1995). Increasing the interval of time between measuring the temperature of the first and the second mouse from 1 to 2, 5, or 10 min shows that the maximal response is always observed in 8 to 10 min (Zethof et al. 1994).

This phenomenon also is known as "psychogenic fever." It is caused by emotional situations and is attributed primarily to brown adipose tissue (BAT) thermogenesis. However, this phenomenon also is observed in adult humans, who have little or no BAT, who anticipate involvement in a sporting event (Renbourn 1960). It is a true fever. It occurs in cold and neutral environments, and can be blocked by sodium salicylate and indomethacin. The latter observation indicates that psychogenic fever is prostaglandin-mediated, but the mechanism by which prostaglandin is suddenly generated remains unclear.

Can We Burn Cancer? Hyperthermia as a Therapeutic Tool

Hyperthermia has been used as a therapeutic approach to the treatment of cancer. Core body temperature is elevated to between 40° and 42°C for hours without causing thermal injury. How is this possible? Why do such patients not suffer heatstroke? What is the role of synthesizing heat shock proteins during such procedures?

The use of heat to treat disease has been practiced for centuries. Recall the beneficial effects of fever discussed in chapter 8. Today, the role of local and whole-body hyperthermia in human cancer therapy is under clinical trial. Hyperthermia was defined in chapter 4 as an elevation in body temperature without a change in set-point. As a method of therapy, hyperthermia may be redefined as an elevation in tissue temperature to 41°C or above in opposition to thermoregulatory control mechanisms. Techniques employed to raise temperature include radiation (electromagnetic, infrared), diathermy, and ultrasound.

The rationale for using hyperthermia in the treatment of malignant disease is based primarily on differences between the vasculature and blood supply in tumors and normal tissue. Tumors frequently have a disorganized and heterogeneous blood supply that can lead to underperfusion and subsequent hypoxia, anaerobic metabolism, and acidosis. Because blood flow to a tissue cools the tissue, tumors or regions of tumors with reduced blood flow will become hotter than normal tissue and more susceptible to killing because of their low pH and limited nutrient supply (figure 9.1). Moreover, heating cells to 42°C–45°C inhibits DNA, RNA, and protein synthesis, and a temperature of 45°C is sufficient to cause direct induction of DNA strand breaks (Streffer 1995). The use of hyperthermia as a treatment modality began in earnest during the 1960s, but the induction and monitoring of hyperthermia is technically challenging. The reader with interest in this area is referred to the text by Field and Hand (1990). The paragraphs below give a brief history of hyperthermia as a therapeutic modality and the use of whole-body heating in the treatment of cancer.

The idea that cancer can be destroyed by a temperature only a few degrees above normal is based on early reports of tumor regression and even cure of patients following infection accompanied by high fever. These observations led Coley (1893) to induce pyrexia in cancer patients by administering bacterial pyrogens. In recent reviews of this approach, it was reported that complete regression and five-year survival occurred in 46 percent of 523 inoperable cases and in 51 percent of 374 operable cases, with better results when higher temperatures were achieved (Nauts 1982a, 1982b).

Today, whole-body hyperthermia usually is achieved by noninvasive techniques such as hot air, hot water, or hot wax, radiation, or a combination of these methods in the anesthetized patient. Invasive techniques include femoral arteriovenous shunt and peritoneal irrigation. The maximal temperature that can be tolerated depends on the thermosensitivity of critical target tissues, such as the heart and lungs, liver, and brain. The maximum is generally assumed to be 42°C, but higher temperatures have been reported (Gerad et al. 1984). Hyperthermia up to 42°C usually does not destroy all tumor cells; as stated earlier, the best results occur when hyperthermia is combined with radiotherapy. An often observed and rather remarkable effect of whole-body hyperthermia is the disappearance of pain immediately after the treatment.

When normal body cells or an entire organism is subjected to a sublethal thermal stress, morbidity and mortality are significantly improved upon exposure to an otherwise lethal thermal stress. This phenomenon is called "thermotolerance" (see chapter 4), and is attributed to the expression of heat shock proteins in virtually all cells that have been tested.

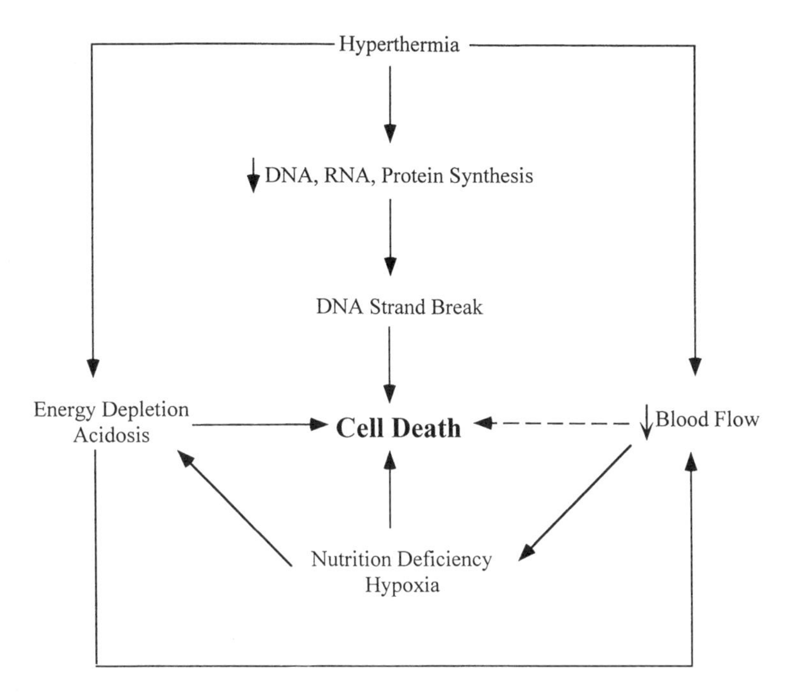

Figure 9.1
Hyperthermia-induced cell death and related physiological and metabolic factors. See also figure 7.4 for cell death mechanisms during an ischemic/hypoxic episode. (Modified from Streffer 1995.)

Whole-body hypertermia can produce thermotolerance depending upon time and temperature of the heat treatment, conditions of the cell (pH, available nutrients, oxygen status), cell cycle, cell type, and cell growth rate. It can increase the slope of a survival curve by a factor of 15 or increase the duration of heating to produce an end point by a factor of 4 or more.

How does hyperthermia produce cell death? Cell inactivation associated with hyperthermia occurs at many levels, including the cell membrane, nucleus, cytoskeleton, and metabolism. There is considerable evidence that hyperthermia modifies the lipid environment of the plasma cell membrane, leading to change in membrane protein structure and function. Whether these changes promote the demise of the cell or produce damage to other cell components is unclear.

Cold Body, Cold Brain: The Mysteries of Hypothermia

In Iowa, during a frigid night in March 1987, a two-year-old toddler, perhaps in his sleep, wandered out of his parents' mobile home. About three hours after his mother found him gone, he was discovered facedown in an ice-covered puddle in a cornfield a half-mile away. His pajamas were frozen to his body and ice had begun to form around his face. He had no heartbeat, he was not breathing, and his body temperature was about 15°C. Clinically speaking, he was dead.

When a police officer came upon the scene, there appeared to be no life in the boy's eyes or face: "he looked like one of those dolls where the eyes roll back in their head" (Ryberg 1987 pp. 1A, 7A). He immediately began cardiopulmonary resuscitation. For 90 minutes there was no response. Then, miraculously, a muscle twitched, a faint heartbeat was felt, and the boy began to breathe on his own. Doctors claimed that a major factor in the boy's favor was that his body temperature was reduced gradually. This lowered his metabolism and reduced the activity of his major organs, thereby lessening the amount of oxygen they needed. Thus, major damage was prevented when his heart stopped. The fact that he stopped breathing the instant his face contacted the water prevented him from drowning in the puddle (Gooden and Elsner 1985). Although he required major physical therapy and had to learn how to walk again, one year later it appeared that the boy had made a complete recovery.

Hypothermia is defined as a reduction in body temperature below 35°C. The rationale for its use is illustrated in figure 9.2. The point to which body temperature can be lowered with survival is not known, but hypothermia can be lethal at any level below 35°C. The young, especially infants, appear to be more tolerant than the elderly. The reason for this in not known, but has been related to the ease with which infants slip into a state of poikilothermy (Blair 1964). Hypothermia is not artificial hibernation, which is impossible in homeotherms. Another form of hypothermia is that produced by drugs. It is called *clinical hypothermia*, and does not involve shivering.

The initial interest in hypothermia during cardiac surgery was fueled by the success experienced with cyanotic infants. Infants are remarkably cold-tolerant. Mothers who are cooled to a moderate level of hypother-

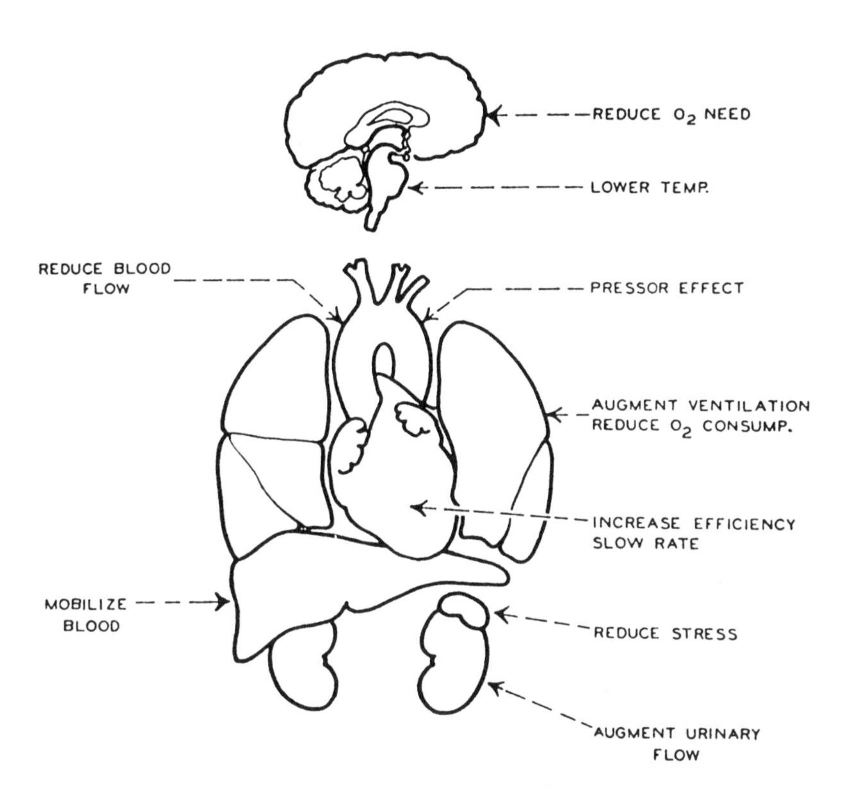

Figure 9.2
The rationale for hypothermia is related to the reduced oxygen requirement (by about one-third), sustained arterial blood pressure produced by the pressor effect of cold, reduced heart rate, and enhanced cardiac efficiency. Augmented breathing improves alveolar ventilation, reflex mechanisms are enhanced, and cerebral inflammatory reactions are reduced. Renal flow may be improved, and blood sequestered in the splanchnic vascular bed may be mobilized. Finally, the stress response of the adrenal cortex may be alleviated. (From Blair 1964.)

mia, 30 to 32°C, deliver normal babies. The rationale for pediatric hypothermia is the relief of the hypoxic state. There is a marked beneficial effect of hypothermia in asphyxia of newborn infants (Blair 1964), probably because of the lower metabolic demands in the infant.

The critical organ in any evaluation of hypothermia is the brain, especially when the circulation is arrested for intracardiac surgery. In a normothermic animal, circulatory arrest for only 6 min produces severe metabolic disturbances due to anoxia. PCO_2 and lactic acid rise and pH falls. In contrast, circulatory arrest for up to 40 min is safe during increasing depths of hypothermia based on clinical, biochemical, and EEG criteria (table 9.1).

Even more important than circulatory arrest, but not equally appreciated, is the effect of hypothermia on reflex mechanisms and the control of homeostasis. As brain temperature descends into a deeper cold state, cortical reflexes are lost first, followed by mesencephalic reflexes, and last by bulbar reflexes. Nerve damage has been reported at 20°C, which has been postulated to be the lethal limit for humans. But is it? Early studies were based on a lack of respiratory or circulatory support. Children have been cooled to below 5°C (Blair 1964), and Niazi and Lewis (1958) cooled a cancer patient to 10°C with successful resuscitation.

Some of the most intriguing experiments with hypothermia have involved supercooling. Hibernating bats have been maintained at –5°C to –7°C for several days. With rewarming, all animals regained activity. However, if crystallization occurred, the bats died. Likewise, deepwater fish swim and feed while in a permanent supercooled state of 1°C; however, if they begin to crystallize and freeze, they die. Even monkeys have been cooled to a temperature just above 0°C with complete recovery, including no gross neurologic abnormalities, after cardiac standstill for 2 h.

There are many aspects of clinical hypothermia that are not understood, and some physicians question its use altogether. However, it has been an effective surgical tool as well as a treatment for cancer and other medical problems. The rationale for its use in cancer therapy was twofold: (a) to relieve pain and (b) to stop growth of the tumor. The latter never materialized, but pain relief was quite successful. Today, hypothermia is used only as an adjunct to either chemotherapy or radiation.

Table 9.1
Safety limits for arrested brain blood flow during increasing levels of hypothermia

°C	37	30	28	25	20	10
°F	99	88	82	77	68	50
Min	3	6	8–10	12–15	20	40

Modified from Blair 1964.

Clearly, it plays an important role in tissue preservation for organ transplantation.

The Sleeping Brain: Hibernation and Dormancy

Hibernation is a unique phenomenon in mammals and birds. Is it different from hypothermia or the dormant state of a reptile? What occurs in the brain of a hibernating animal? What advantage was given when nature evolved such a process? Does hibernation have survival value? If it does, why has evolution preserved this phenomenon in some mammals, but not in those that represent the main line of evolution? What allows a ground squirrel to lower its body temperature more than 30°C and remain without food for three to four months, except for a few periods of arousal? What mechanisms allow a bear to maintain a state of dormancy for more than 6 months without food and water? Could hibernation or dormancy serve as a useful tool during travel through space?

Hibernation in mammals is characterized by a change in body temperature of more than 10°C that is produced by a decrease in metabolic rate. It is not simply a state of hypothermia, because a hibernating animal retains the capacity to rewarm spontaneously to a normothermic level without absorbing heat from the environment. An animal that is hypothermic can not raise body temperature by metabolic means. Herein lies the difference between the dormant state of the reptile and the hibernating state of the mammal. The reptile lacks the capacity to warm itself. As stated earier, if left in a cold environment, it eventually will die. On the other hand, mammals and birds will spontaneously rewarm in response to an internal or external signal. Internal signals might include some annual

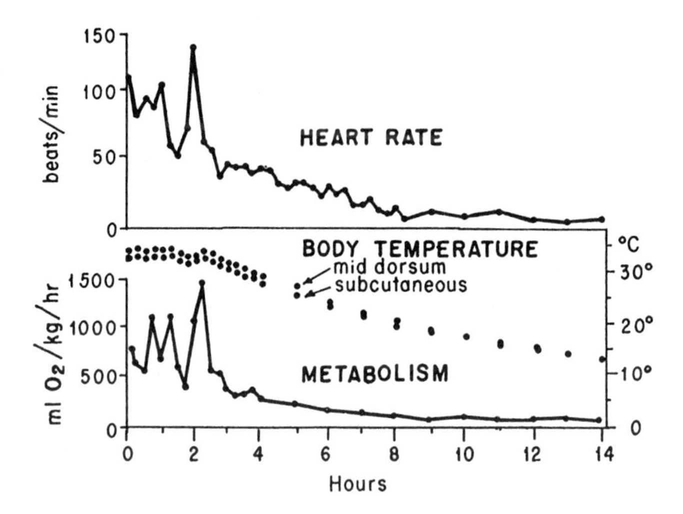

Figure 9.3
Woodchuck entering dormancy. Note the dissociation of physiological responses. Heart rate drops, first followed by oxygen uptake and then body temperature. (Taken from Lyman 1958.)

rhythm or lack of REM sleep. External signals might include vibration or a slight change in ambient temperature.

Hibernation clearly has survival value because it represents the ability of an animal to sustain life in times of limited availability of food and water. Evidence to support this concept is that over several days dormant pocket mice lose less weight than nondormant, unfed controls (Bartholomew and Cade 1957). Why has it not been retained in primates and humans? It seems that nature changed strategies during evolution. In some species of animals, constant and increased wakefulness was deemed more valuable than hibernation, perhaps because of predators, extreme temperatures, or abundant food and water throughout the year.

What role does the brain play in hibernation? After the cold stimulus, is the brain the initial mover, or does it passively follow a general body state initiated by the endocrine system? Hibernation begins with a decrease in heart rate followed by a decrease in metabolism (20- to 100-fold) and respiratory rate, and finally by a decrease in body temperature (figure 9.3). Brain temperature drops slowly, 2°C to 4°C/h. Hormone (insulin, glucagon, thyroxin) secretion declines, in part from a decrease in hypophyseal activity, and cardiovascular function slows. For example, the heart rate of ground squirrels decreases from a mean of around 300 beats/min at rest to 7–10 beats/min in the hibernating animal. Breathing decreases accordingly, and enzyme activity slows the time of metabolic reactions.

Is the end result of this form of cataleptic state a "sleeping brain"? Certainly not! Spontaneous cortical activity can still be recorded in some animals, although not in all, and cortical activity can still be elicited following peripheral nerve stimulation. The fact that the heart continues to beat and the animal continues to respire indicates that the autonomic nervous system remains active. The reticular system of the midbrain is the area least resistant to the cold, which is in accordance with the physiological responses observed in the hibernating period. The animal enters into a stage of NREM sleep, as it normally does every day when it goes to sleep. However, during the onset of hibernation, NREM sleep is prolonged without its entering REM sleep. It looks as if REM sleep is selectively suppressed (Walker et al. 1977). (This interesting observation indicates that the animal is still thermoregulating. During REM sleep [see chapter 4] mammals lose their capacity to thermoregulate, and they become poikilothermic.) Below a body temperature of 25°C, cortical activity is almost suppressed. The EEG becomes isoelectric at a body temperature of about 20°C. EEG recordings in different structures of the brain have shown that after the cortical EEG becomes quiescent, the thalamus, then the hippocampus, become quiescent (Mihailovic 1972). In contrast, the hypothalamus remains active, with intermittent increases in activity from other areas of the limbic system (Strumwasser 1959). (This is another interesting observation compatible with the idea that a hibernating animal is constantly controlling its body temperature by the brain.) During arousal from hibernation, a similar sequence, in the reverse order, occurs.

Are entry into hibernation and arousal from hibernation active or passive processes? To address this question, neuronal activity in the brain was studied during different phases of hibernation, using ^{14}C–2-deoxyglucose (Kilduff et al. 1990). Relative uptake of 2-deoxyglucose by brain cells was determined for 96 brain regions during 7 phases of the hibernation cycle: (a) euthermia; (b) 3 body temperature intervals entering hibernation; (c) stable, deep hibernation; and (d) 2 phases of body temperature during arousal (figure 9.4). Statistical analysis selected brain sites that showed a similar pattern across all phases of hibernation. Three factors were identified that accurately discriminated among these various phases. Factor 1 (the hypothalamic factor) was associated with brain regions that showed a trend for increased 2-deoxyglucose uptake as a function of the depth of hibernation. Factor 2 (the cortical factor) was characterized by regions showing a rapid decrease in 2-deoxyglucose uptake early in hibernation; and factor 3 (the raphe factor) was characterized by regions exhibiting a pattern including factors 1 and 2. Thirty-three brain regions exhibited a relatively high correlation on factor 1, including almost all hypothalamic regions. To show the relationship between the factor score for factor 1 and 2-deoxyglucose for a particular brain region, figure 9.4B shows the 2-deoxyglucose profile for the suprachiasmatic nucleus, a hypothalamic structure that reflected a pattern of 2-deoxyglucose that was inversely associated with the depth of hibernation. Figure 9.4C shows the 2-deoxyglucose profile for the cingulate cortex, a brain region that reflected a pattern of rapidly decreasing 2-deoxyglucose early in entrance into hibernation.

The analysis revealed that neuronal activity in the cerebral cortex decreased, and hypothalamic activity increased, during entrance into hibernation, whereas during arousal the opposite occurred. The suprachiasmatic nuclei were the main hypothalamic nuclei participating in the entrance into hibernation, and the cingulate cortex was a major cortical area that became active in arousal.

Disinhibition of structures such as the frontal and cingulate cortexes, which are mainly inhibitory to limbic structures, seems to play a crucial role in this initial phase of hibernation. As a consequence, the limbic cortex is highly sensitive to stressful environmental stimuli (Gabriel et al. 1977). Arousal from hibernation seems to be triggered by a different pattern of neuronal activity. In this case, suprachiasmatic and paraventricular nuclei of the hypothalamus, as well as the medial preoptic area, play major roles (figure 9.5). These results support the hypothesis that hibernation is an active orchestration of integrated neurophysiological events rather than a passive process (Kilduff et al. 1990).

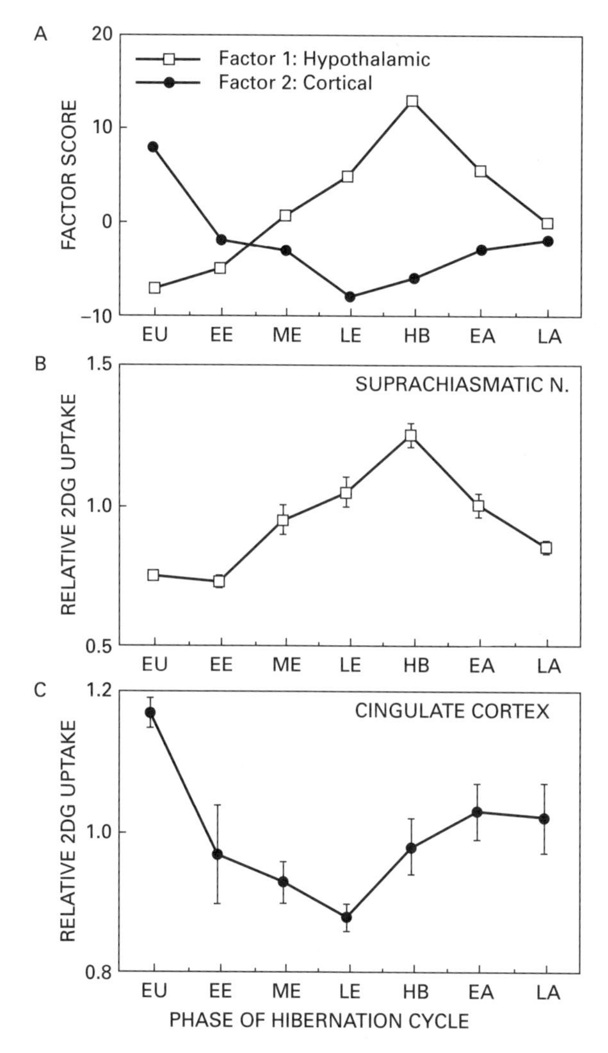

Figure 9.4
A. Factor scores for factor 1 (hypothalamic) and factor 2 (cortical) across the phases of the hibernation cycle. B. Relative 2-deoxyglucose uptake (mean ± SEM) across the hibernation cycle for the suprachiasmatic nucleus, a representative brain region that loaded (i.e., correlated with) strongly on factor 1. C. Relative 2-deoxcyglucose uptake (mean ± SEM) across the hibernation cycle for the cingulate cortex, a representative brain region that loaded strongly on factor 2. EU, euthermia; EE, early entrance; ME, midentrance; LE, late entrance; HB, hibernation; EA, early arousal; LA, late arousal. (Taken from Kilduff et al. 1990.)

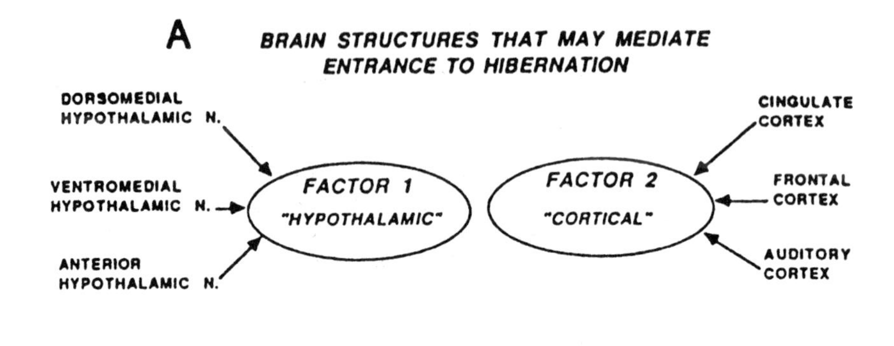

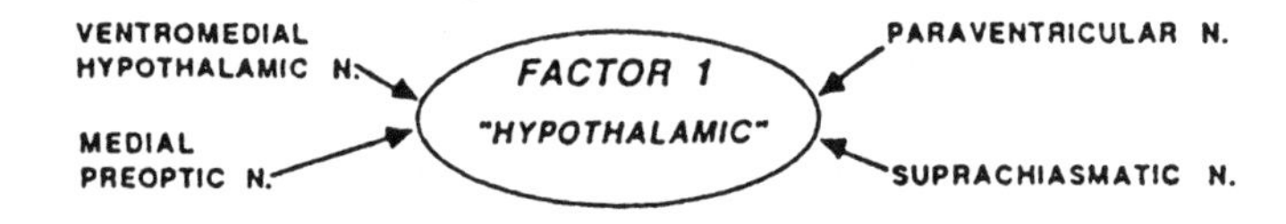

Figure 9.5
(A) Schematic diagram illustrating hypothalamic and cortical involvement in the entrance into hibernation. (B) Schematic diagram illustrating hypothalamic involvement in the arousal from hibernation. (Taken from Kilduff et al. 1990.)

A Way in Space: Hypothermia or Hibernation?

Hypothermia

Many fantasies and literary tales have reported the revival of people after they were frozen solid. For example, the imagination of many was stimulated by the tale of a French soldier who, during the retreat of Napoleon's forces from Russia, became frozen and was revived years later without damage to his body except the accidental loss of an ear during his reanimation. But true stories like that of Johnny Stevens, the young child who recovered completely after having a body temperature of 17.8°C and a respiratory rate of 3 breaths/min, have made some people rethink the possibility of cooling and reviving a human. An obvious application of such a procedure would be traveling through space. However, despite isolated cases like the one described above, the adult human body cannot be cooled below 25°C (Folk 1974: 212). Therefore, the idea of freezing a human for prolonged travel in space remains in the realm of science fiction.

Hibernation

If hypothermia is not feasible as a physiological state to travel in space, what about the possibility of hibernation? There is no doubt that the possibility of hibernating, for at least part of such a long journey, without consuming food and water or having waste products, could be of considerable advantage. In previous paragraphs an orchestrated pattern of brain activity was turned on when an animal entered and emerged from hibernation. Thus, two intriguing questions are (a) Could a selective convergent pattern of artificial external stimulation (electrical, electromagnetic fields) in different nuclei and regions of the brain, together with a concomitant pattern of lower environmental temperature, trigger hibernation in humans? (b) Could a selective convergent pattern of brain stimulation be programmed to trigger arousal mechanisms after a certain period of time? Could a research program conducted in this direction, in nonhibernating mammals, be of interest?

Is there any real possibility of designing a "hibernation pill" that would lower metabolic rate without long-term detrimental effects on primary physiological functions, yet provide the advantage of reducing food and water requirements? The answer is "not in the immediate future." Evidence for a chemical that can induce hibernation has been reviewed by Folk (1974). The substance, called "hibernation trigger," has been discovered in the blood of hibernating ground squirrels and woodchucks. It is absent from the blood of nonhibernating animals and induces hibernation in ground squirrels during the summer by lowering metabolic rate. The idea of injecting such a substance into an astronaut on a prolonged journey into space is appealing, but many more experiments are needed before serious consideration can be given to such an idea.

10

The Brain Is the Body: A Unitary Perspective on Thermoregulation

The Brain, the Body, and Keeping Warm

For more than 500 million years the function of the brain has been coded with *time*. The brain has served the rest of the body by responding to changes in the environment, such as temperature, within a minimum time. Survival of the species depended on the time of response to the stimulus produced from a relatively well defined environment to which animals were well adapted. There was limited capacity in the brain for flexible behavior. In other words, before responding to any noxious (predator or poison) or pleasurable (food, water or sex) stimulus, there was little time to think. For lower vertebrates (fish, frogs, and reptiles) to develop and perform survival behaviors such as seeking food and water, and mating, warming of the brain from external sources was essential. Brain activity, and hence body activity (behavior), was impossible without the brain and body achieving a certain range of temperature. In this case, and since the beginning of life, radiation from the sun was the main source of heat. Thus, the primary event required for life, even before acquiring food and water, was generating a *hot brain*.

With the appearance of mammals, the brain acquired greater capacity for processing information from the environment and was able to provide a more effective response to a determined stimulus. Evolution changed strategies. With mammals, the relatively fixed behavioral responses that served survival so well for millions of years came to an end. And, as is so often the case, an end was only a new beginning.

The Dawn of the Warm Brain and Intelligence

Primitive mammals acquired bigger brains than their ancestors and contemporaries, the reptiles (Jerison 1973). Their brains acquired circuits to regulate body temperature at an elevated level, thus facilitating their movement in and out of different thermal environments. Cold no longer impaired their alertness, readiness to escape predators, or their ability to seek food and water. The brain was *maintained* at a *warm* temperature. An intriguing question is "What came first, the acquisition of a bigger brain, and as a consequence the acquisition of endothermy, or the reverse, the acquisition of endothermy followed by enlargement of the brain? It is tempting to speculate on the latter possibility.

It appears that mammal-like reptiles that entered the new environment (nocturnal niches) lacking radiant heat from the sun were already equipped with the anatomical machinery to generate a certain amount of heat in their own bodies (chapters 2 and 3). As described in chapter 2, it is likely that a certain reorganization of the existing neural tissue around the hypothalamus (posterior hypothalamus?) and the rest of the limbic system may have provided the control of input-output information for the regulation of body temperature. In the almost complete absence of radiant heat from the sun, nights for these creatures must have been extremely cold, and challenging to the point of imperiling survival. It is possible that initially small changes occurred in the brains of some of these creatures so that without a significant increase in the total amount of brain tissue (reorganization), their descendants were more capable of survival under these conditions.

With the acquisition of a certain degree of endothermy (see chapter 2), it is more likely, to think in terms of Jerison's hypothesis, that selection pressure toward the acquisition of a bigger brain was effective. In fact, with a constant and active warm brain, the trend toward an increase in brain tissue incorporating new circuits to enable hearing, smell, and sight in the darkness was more conceivable (Jerison 1973).

Did a Warm Brain Speed the Pace of Evolution?

To be warm is an inherent property of life. To be *constantly* warm was probably a key step in evolution that led to the ultimate achievement in development, the human organism. Creatures with a warm and perpetually active brain throughout the year had an internal environment that allowed for constant trial-and-error learning and an enduring opportunity for mutations to flourish so that accelerated evolution occurred.

Life on Earth appeared about 3 billion years ago. Unicellular existence lasted for more than 2 billion years before multicellular organisms appeared. With the development of multicellular organisms, evolution accelerated. The Paleozoic era, characterized by the appearance of invertebrates, fish, amphibians, and primitive reptiles, lasted for only about 325 million years, followed by the Mesozoic era (the age of reptiles) and the appearance of the first and most archaic mammals (and with them the process of endothermy). The Mesozoic era lasted for only 180 million years. With the Cenozoic era (the age of mammals), a tremendous acceleration of evolution took place that lasted for only 65 million years. In this latter period humans appeared.

And what can be said about the brain? With the appearance of mammals, there was an acceleration in the evolution of the brain. From australopithecines to modern man, the brain increased in weight by almost 1 kg in just 2–4 million years. There is no doubt that the increase in brain size and brain organization must have been the product of multifactorial convergent events. It would be simplistic to believe that brain evolution could be attributed to a single event, such as assuming an upright posture, toolmaking, the acquisition of language, or social modes of life. In this very short period of time, 2 million years, brain weight increased from 600–700 in *Homo habilis* to 900–1,000 grams in *Homo erectus,* to 1,300 grams in *Homo sapiens,* and finally to 1,400 grams in *Homo sapiens sapiens* (Tobias 1971; Conroy et al. 1998). There must be, together with the factors already mentioned, unknown factors, some perhaps more important than others, that influenced humans in their acquisition of a larger brain. And most important, there must have been key factors responsible for triggering this endurance race toward the acquisition of a bigger brain. Could environmental temperature, together with primitive hunting behavior in early hominids, be one of these key factors?

Was Hunting Behavior the Engine of Brain-Mind Processes?: The Hypothetical Role of Memory

Krantz (1968) suggested that hunting behavior in its more primitive form, "persistence hunting," could have been the basis of the selective pressures that paved the evolutionary path for australopithecines to acquire bigger brains. Persistence hunting, as Krantz describes it, is a predatory behavior unique to humans. Primitive people who practice this kind of hunting behavior today include the Shoshonean Indians of North American, the Tarahumara Indians of Mexico, and the Kalahari Bushmen of Africa. The essence of this type of hunting is persisting in the chase of the prey for as long as one or two days, which obviously requires the hunter to focus constantly on the task at hand. This is a kind of brain-mind training that clearly could be relevant for the development of the brain. That the australopithecines could initially be primitive hunters is a strong possibility, if we take into account the possibility that they were creatures better adapted to running than to walking. This is consistent with early suggestions that the transition from ape to human was predatory (Darwin 1871; Dart 1964).

The idea that this kind of primitive hunting could be relevant for developing bigger brains within a species is based on the assumption that individuals who were anatomically adapted to run and who also had bigger brains would have better memories than individuals with smaller brains. This clearly would have important survival value. Krantz describes his ideas in the following terms:

> The idea of persistence hunting permits the following hypothesis as to the selective forces that brought about the transformation from Australopithecus to Homo: small steps in the enlargement of the Australopithecus brain would have been of selective advantage mainly by increasing the time and distance that the possessor would be able to pursue his mobile food supply. Considering the young, injured and aged as well as normal adults of all species of potential food available to our ancestors, there was a continuous gradation in pursuit times necessary to bring down game. At first, Australopithecus could run down only those animals most quickly exhausted, and must have been in keen competition with many other carnivores. As the reward in food for successful pursuit of game tended, on the average, to go to those individuals with the greater mental time spans, selective pressure would favor larger brains with better memories. (Krantz 1968: 451)

Did Heat Stress Contribute to the Acquisition of a Bigger Brain? Fialkowski's Hypothesis

Based on Krantz's hypothesis of primitive *persistence hunting* in australopithecines and assuming that the initial process of hominization occurred in the tropics (southern and eastern Africa), Fialkowski (1986) hypothesized that initially the australopithecines hunted by using individual techniques involving constant, persistent, and long running pursuits of prey through the savannas, before developing the well-organized cooperative hunting techniques attributed first to *Homo erectus* (Tobias 1971). It is in this context that Fialkowski argued that "the loss of fur, one of the most conspicuous evolutionary changes which occurred in the homind line, was primarily a result of the adaptation to a greatly increased heat stress associated with this type of behavior" (1986: 288). Could an increase in body temperature affect the functioning and evolution of the brain?

Fialkowski (1986) points out the difference in the mechanisms by which prehuman hunters and their prey adapted to thermal stress. As discussed at length in chapter 6, fast-running mammals in the savannas developed "selective brain cooling" as a mechanism to prevent overheating of the brain. For example, in a small antelope, brain temperature during high-speed running for 7 min was almost 3°C below body temperature (Taylor and Lyman 1972). This is attributed to the presence of an intracranial heat exchanger, the carotid rete, that allows hot carotid blood from the trunk to be cooled by venous blood draining the nasal and oral cavities. In contrast, prehuman hunters were poorly adapted for heat stress and, as a result, probably developed very high brain temperatures that compromised their hunting capacity. To cope with this dilemma, Fialkowski (1978, 1986) proposes that the size of the brain increased. The reasoning behind this strategy is based on the ideas of von Neumann (1963), a mathematician and computer scientist. The scenario is as follows. Prehuman hunters began to develop a generalized cooling system, that is, to lose their dense fur; but long before this process was complete, they were forced to perform long runs (which could not be interrupted or abandoned, since that resulted in hunting failure) without possessing a fully developed mechanism of cooling through sweat evaporation. Moreover, cooling by evaporation required a high ingestion of water, and after a long run at high temperature, sweating could decrease or, in the case of high humidity, be less effective than in dry air. In any case, blood temperature could rise high enough to limit the function of brain cells.

According to von Neumann (1963), it is possible to obtain a reliable system (brain and heat) even if its elements are working unreliably (neurons affected by heat), provided there are sufficient elements (neurons) and interconnections between the elements. Fialkowski applies this concept to the problem of overheating in prehuman hunters. He surmises that even though inadequate cooling causes some brain elements to lose their reliability, a greater number of neurons and interconnections between neurons as a result of a larger brain would allow the brain, as a whole, to continue functioning. In this way, Fialkowski proposes, early hominds were under very strong selective pressure to acquire bigger brains that were more heat-resistant. He states that "hunters more susceptible to a cognitive crisis during hunting raids would have higher average mortality (failure in hunting, failure to escape predators)" (Fialkowski 1978: 90). In essence, this statement is analogous to Krantz's hypothesis regarding memory: individuals with bigger brains would have more heat-resistant brains than individuals with smaller brains.

Heat, Thirst, Memory, and the Human Brain

Eckhardt (1987) has criticized Fialkowski's idea that the reliability of a complex system such as a computer could be enhanced by increasing the number of elements and the interactions among them. He maintains that increasing the number of processing elements (chips or neurons) does not increase the reliability of the system; rather, it has just the opposite effect. In fact, he maintains that "the existing solution to this problem in computer science is not merely to add more chips, but to integrate an error-correcting code. This strategy which enables the memory to function even after hundreds of errors (typically accumulated over many years) is perhaps more analogous to neural reorganization than to memory expansion via mere repetitive subunit addition" (1987: 194). If the brain increased the number of neurons under heat stress to cope more effectively with a hot environment, one would expect this increase to occur in the part of the brain responsible for vital functions—more specifically, the areas involved in the control of temperature regulation, such as the hypothalamus. But the fact is that the areas of the brain which have increased in size over the last 2–4 million years are the association areas of the cortex, which apparently have nothing to do with vital functions such as temperature regulation. To that Fialkowski responds, "It is possible that the cerebral association areas have changed their functions, originally having been mainly oriented toward the vital ones" (1986: 290).

It is interesting that Eckhardt, who so strongly criticized Fialkowsi, has admitted that some relationship can exist between heat stress and an increase in brain size among early hominids. In fact, he suggests that the relationship between heat stress and increasing hominid brain size was not the result of heightened resistance to heat-induced brain malfunctions but, rather of an increased capacity or duration of information storage.

Thus, once again, memory comes into play. Recall Krantz's hypothesis that memory was probably one of the more powerful selective pressures favoring enlargement of the brain. In this case, Eckhardt suggests that in situations where early hominids had to survive in the hot, dry savanna, memory of water holes in areas where they hunted could have had survival value. Present-day Bushmen hunt an area as large as 10,000 km^2 and from memory know the location of every water hole: "no other predator or prey species matches that level of information storage and retrieval" (Eckhardt 1987: 199).

Therefore, heat and memory may be two of several factors that converged on the force that increased selective pressures toward the acquisition of a bigger brain. If this were the case, then we would be forced to admit that toolmaking, new social modes of life, and so on would have been consequences rather than causes of the initial increase in brain size during evolution. That is, in fact, Fialkowski's interpretation. To paraphrase his conclusion, the increase in size and complexity of the hominid brain was largely a side effect of a quite different adaptation that had little to do with abstract thinking and complex reasoning. It was, instead, an evolutionary response to greatly increased heat stress under conditions of primitive hunting (Fialkowski 1978, 1986).

The Warm Bed of Our Mental Processes

Is the mind (mental processes) a "side effect" function of a brain developed through evolution just for survival purposes under selective, although multifactorial, environmental pressures? This is one of the most intriguing and crucial questions in philosophical anthropology. Evolutionary biology is providing pieces of a complicated puzzle that we hope will lead to a better understanding of the human brain. What seems more and more evident from different scientific approaches to the study of humans, and in a more consistent way, is that "Human brains evolved from the brains of preceding animals, sharing much with them structurally and functionally as well as cognitively. However remarkable the human brain is, it is a product of Darwinian evolution, with all the constraints that such a history implies" (Llinas and Churchland 1996: ix). Given this perspective, environmental temperature was probably one of the multiple factors that contributed to the development of humans through evolution.

This suggests that body, brain, and mind are a continuum under a unitary construction of humans through time. (Mora, 1995, 1999) In this process, there could have been circumstances or needs that led to the development of anatomical systems designed originally to serve one function but, over time and through exposure to different stresses, were modified or used to serve other functions. This could have been the case with the brain and the appearance of mind processes. If we assume for a minute that Fialkowski's hypothesis is correct, the enlargement of the brain initially evolved to cope with environmental heat stress until more effective heat-dissipating systems evolved. When the loss of fur was complete and eccrine sweat glands evolved over the general body surface to increase evaporative cooling dramatically, more of the brain became available for cognitive functions. In turn, these cognitive functions proved to have survival value and have been used to preserve the integrity of humans through a constant flow of information from the environment to the body-brain-mind and from them to the environment in a more and more efficient manner.

Body-brain-mind-environment represents a flow of functional information. Modern biology is clearly showing that there is a continuum between biochemistry-morphology and function, particularly with regard to the brain. Thus, changes in the environment induce neural activity (release of neurotransmitters) that could induce new synthesis of proteins—the basis of learning and memory processes (Kandel et al. 1991). New proteins change the morphology of neural systems, which in turn changes the function of the system. These plastic changes within the brain, induced by changes in our sensory environment and also in our body, are the basis of learning and memory processes. These changes provide a constant flow of information that ultimately leads to an adaptation between body-brain and mind. It is in this way that "our growing sense of whatever the world outside may be, is apprehended as a modification in the neural space in which body and brain interact. It is not only the separation between mind and brain that is mythical: the separation between mind and body is probably just as fictional. The mind is embodied, in the full sense of the term, not just embrained" (Damasio 1994: 117–118). These reflections are of paramount importance for an understanding of human nature (Mora 1996).

References

Adams, W. C., R. H. Fox, et al. (1975). Thermoregulation during marathon running in cool, moderate, and hot environments. *J. Appl. Physiol.* 38: 1030–1037.

Adamson, K. J., G. M. Gandy, et al. (1965). The influence of thermal factors upon oxygen consumption of the newborn human infant. *J. Pediatrics* 66: 495–508.

Adolph, E. F. (1947). Tolerance to heat and dehydration in several species of mammals. *Am. J. Physiol.* 151: 564–575.

Adolph, E. F., and W. B. Fulton. (1923–1924). The effects of exposure to high temperatures upon the circulation in man. *Am. J. Physiol.* 67: 573–588.

Alan, J. R., and C. G. Wilson. (1971). Influence of acclimatization on sweat sodium concentration. *J. Appl. Physiol.* 30: 708–712.

Alexander, D. P., K. E. Cooper, et al. (1989). Sodium salicylate: Alternate mechanism of central antipyretic action in the rat. *Pflugers Arch.* 41: 451–455.

Almirall, H., A. Aguirre, et al. (1993). Temperature drop and sleep: Testing the contribution of SWS in keeping cool. *NeuroReport* 5: 177–180.

Arjona, A. A., D. M. Denbow, and W. D. Weaver, Jr. (1990). Neonatally-induced thermotolerance physiological responses. *Comp. Biochem. Physiol.* 95A: 393–399.

Armstrong, L. E., J. P. De Luca, and R. W. Hubbard. (1990). Time course of recovery and heat acclimation ability of prior exertional heatstroke patients. *Med. Sci. Sports Exerc.* 22: 36–48.

Aschoff, J. (1965). Circadian rhythms in man. *Science* 148: 1427–1432.

Atkins, E. (1991). Foreward. In *Fever: Basic Mechanisms and Management,* vii–viii. New York: Raven Press.

Atkins, E., and P. Bodel. (1972). "Fever." *N. Eng. J. Med.* 286(1): 27–34.

Austin, M. G., and J. W. Berry. (1956). Observations on one hundred cases of heatstroke. *JAMA* 161: 1525–1529.

Bach, V., F. Telliez, et al. (1996). Body temperature regulation in the newborn infant: Interactions with sleep and clinical implications. *Neurophysiol. Clin.* 26: 379–402.

Baker, M. A. (1982). Brain cooling in endotherms in heat and exercise. *Ann. Rev. Physiol.* 44: 85–96.

Baker, M. A. (1993). A wonderful safety net for mammals. *Nat. Hist.* 8: 63–64.

Baker, M. A., and M. J. M. Nijland. (1993). Selective brain cooling in goats: Effects of exercise and dehydration. *J. Physiol.* (London) 471: 679–692.

Bakker, R. T., ed. (1986). *The Dinosaur Heresies.* New York: William Morrow.

Barrett, J., M. Morris, et al. (1987). The sleep-evoked decrease of body temperature. *Sleep Res.* 16: 596.

Barrick, R. E. (1994). Thermal physiology of the dinosauria: Evidence from oxygen isotopes in bone phosphate. *DINO FEST.* 243–254.

Bartholomew, G. A., and T. J. Cade. (1957). Temperature regulation, hibernation and aestivation in the little pocket mouse. *J. Mammal.* 38: 60–72.

Baum, E., K. Bruck, et al. (1976). Adaptive modifications in the thermoregulatory system of long-distance runners. *J. Appl. Physiol.* 403: 404–410.

Bell, R. C., and J. M. Lipton. (1987). Pulsatile release of antipyretic neuropeptide α-MSH from septum of rabbit during fever. *Am. J. Physiol.* 252: R1152–R1157.

Bennett, I. L., and A. Nicastri. (1960). Fever as a mechanism of resistance. *Bacteriol. Rev.* 24: 16–34.

Benzinger, T. H., C. Kitzenger, et al. (1963). The human thermostat. In *Temperature: Its Measurement and Control in Science and Industry,* J. D. Hardy (ed.), 637–665. New York: Reinhold.

Bird, A. R., K. D. Chandler, et al. (1981). Effects of exercise and plane of nutrition on nutrient utilization by the hind limb of the sheep. *Aust. J. Biol. Sci.* 34: 541–550.

Blair, E. (1964). *Clinical Hypothermia.* New York, McGraw-Hill.

Blatteis, C. M. (1981). The newer putative central neurotransmitters: Roles in thermoregulation. *Fed. Proc.* 40: 2735–2740.

Blatteis, C. M. (1984). *Endogenous Pyrogen: Fever and Associated Effects.* New York: Raven Press.

Blatteis, C. M., and E. Sehic. (1997). Fever: How may circulating pyrogens signal the brain? *NIPS* 12: 1–9.

Bligh, J. (1966). The thermosensitivity of the hypothalamus and thermoregulation in mammals. *Biol. Rev.* 41(3): 317–368.

Bligh, J. (1972). Neuronal models of mammalian temperature regulation. In *Essays on Temperature Regulation,* J. Bligh and R. E. Moore (eds.), 105. Amsterdam: Elsevier Science.

Bosenberg, A. T., J. G. Brock-Utne, et al. (1988). Strenuous exercise causes systemic endotoxemia. *J. Appl. Physiol.* 65(1): 106–108.

Boulant, J. (1980). Hypothalamic control of thermoregulation. In *Handbook of the Hypothalamus,* P. J. Morgane and J. Panksepp (eds.), 1–82. New York: Marcel Dekker.

Boulant, J. A. (1996). Hypothalamic neurons regulating body temperature. In *Handbook of Physiology,* Section 4: *Environmental Physiology,* M. J. Fregly and C. M. Blatteis (eds.), 105 (American Physiological Society). New York: Oxford University Press.

Boulant, J. A., and N. L. Silva. (1989). Multisensory hypothalamic neurons may explain interactions among regulatory systems. *NIPS* 4 (December): 245–248.

Brengelmann, G. L. (1987). Dilemma of body temperature measurement. In *Man in Stressful Environments: Thermal and Work Physiology,* K. Shiraki and M. K. Yousef (eds.), 5–22. Springfield, IL: C. Thomas.

Brengelmann, G. L. (1993). Specialized brain cooling in humans. *FASEB J.* 7(12): 1148–1152.

Brezina, V., and K. R. Weiss. (1997). Analyzing the functional consequences of transmitter complexity. *Trends Neurosci.* 20: 538–543.

Brinnel, H., M. Cabanac, et al. (1987). Critical upper levels of the body temperature, tissue thermosensitivity and selective brain cooling in hyperthermia. In *Heat Stress: Physical Exertion and Environment,* J. R. S. Hales and D.A. B. Richards (eds.), 209–240. Amsterdam: Elsevier Science.

Britt, B. A. (1996). *Malignant Hyperthermia.* New York: Academic Press.

Brock-Utne, J. G., S. L. Gaffin, et al. (1988). Endotoxaemia in exhausted runners following a long-distance race. *S. Afr. Med. J.* 73: 533–536.

Brown, S. J., C. V. Gisolfi, et al. (1982). Temperature regulation and dopaminergic systems in the brain: Does the substantia nigra play a role? *Brain Res.* 234: 275–286.

Bruck, K., and H. Olschewski. (1987). Body temperature related factors diminishing the drive to exercise. *Can. J. Physiol. Pharmacol.* 65: 1274–1280.

Bruck, K., and E. Zeisberger. (1990). Adaptive changes in thermoregulation and their neuropharmacological basis. In *Thermoregulation: Physiology and Biochemistry,* E. Schonbaum and P. Lomax (eds.), 255–307. New York: Pergamon Press. 255–307.

Burger, F. J., and F. A. Fuhrman. (1964). Evidence of injury by heat in mammalian tissues. *Am. J. Physiol.* 206: 1057–1061.

Burton, A. C. and Edholm, O. G. (1955) Man in a cold environment. Williams and Wilkins. Baltimore 1955

Cabanac, M. (1986). Keeping a cool head. *NIPS* 1: 41–44.

Cabanac, M. (1993). Selective brain cooling in humans: "Fancy" or fact? *FASEB J.* 7: 1143–1147.

Carafoli, E. (1987). Intracellular calcium homeostasis *Ann. Rev. Biochem.* 56: 395–433.

Carlisle, H. J. (1969). The effects of preoptic and anterior hypothalamic lesions on behavioral thermoregulation in the cold. *J. Comp. Physiol. Psychol.* 69: 391–402.

Chambers, W. W., M. S. Seigel, et al. (1974). Thermoregulatory responses of decerebrate and spinal cats. *Exp. Neurol.* 42: 282–299.

Cheng, C., T. Matsukawa, et al. (1995). Increasing mean skin temperature linearly reduces the core-temperature thresholds for vasoconstriction and shivering in humans. *Anesthesiology* 82: 1160–1168.

Christ, J. F. (1969). Derivation and boundaries of the hypothalamus, with atlas of hypothalmic grisea. In *The Hypothalamus,* W. Haymaker, E. Anderson, and W. J. H. Nauta (eds.), 13–60. Philadelphia: Thomas Books.

Christman, J. V., and C. V. Gisolfi. (1985). Heat acclimation: Role of norephinephrine in the anterior hypothalamus. *J. Appl. Physiol.* 58: 1923–1928.

Clark, W. G. (1979). Changes in body temperature after administration of amino acids, peptides, dopamine, neuroleptics and related agents. *Neurosci. Biobehav. Rev.* 3: 179–231.

Clark, W. G., and M. J. Fregly. (1996). Evidence for roles of brain peptides in thermoregulation. In *Handbook of Physiology,* Section 4: *Environmental Physiology,* M. J. Fregly and C. M. Beatteis (eds.), 139–153 (American Physiological Society). New York: Oxford University Press.

Coley, W. B. (1893). The treatment of malignant tumors by repeated inoculations of erysipelas: With a report of ten original cases. *Am. J. Med. Sci.* 105: 486–511.

Conroy, G. C., G. W. Weber, et al. (1998). Endocranial capacity in an early hominid cranium from Sterkfontein, South Africa. *Science* 280: 1730–1731.

Cooper, J. R., F. E. Bloom, and R. H. Roth. (1991). *The Biochemical Basis of Neuropharmacology.* 6th ed. New York: Oxford University Press.

Corbit, J. D. (1969). Behavioral regulation of hypothalamic temperature. *Science* 166: 256–257.

Coridis, D. T., R. B. Reinhold, et al. (1972). Endotoxaemia in man. *Lancet* 1: 1381–1386.

Cowles, R. B. (1946). Fur or feathers: A result of high temperatures? *Science* 103: 74–75.

Craig, E. A., and C. A. Gross. (1991). Is hsp70 the cellular thermometer? *TIBS* 16: 135–140.

Crawshaw, L. I., and H. T. Hammel. (1971). Behavioral thermoregulation in two species of antarctic fish. *Life Sci.* 10(1): 1009–1020.

Crawshaw, L. I., and H. T. Hammel. (1974). Behavioral regulation of internal temperature in the brown bullhead, Ictalucrus nebulosus. *Comp. Biochem. Physiol.* A47: 51–60.

Crawshaw, L. I., B. P. Moffitt, et al. (1981). The evolutionary development of vertebrate thermoregulation. *Am. Scientist* 69: 543–550.

Crawshaw, L., L. Wollmuth, et al. (1990). Body temperature regulation in vertebrates: Comparative aspects and neuronal elements. In *Thermoregulation: Physiology and Biochemistry,* E. Schonbaum and P. Lomax (eds.), 209–220. New York: Pergamon Press.

Crompton, A. W., C. R. Taylor, et al. (1978). Evolution of homeothermy in mammals. *Nature* 272: 333–336.

Crosby, E. C., and M. J. C. Showers. (1969). Comparative anatomy of the preoptic and hypothalmic areas. In *The Hypothalamus,* W. Haymaker, E. Anderson, and W. J. H. Nauta (eds.), 61–135. Philadelphia: Thomas Books.

Crosby, E. C., and R. T. Woodburne. (1940). The comparative anatomy of the preoptic area and the hypothalamus. In *The hypothalamus and Central Levels of Autonomic Function,* J. F. Fulton, S. W. Ranson, and A. M. Frantz (eds.). Baltimore: Williams and Wilkins.

Dahlstrom, A., and F. Fuxe (1964). Evidence for the existence of monoamine containing neurons in the central nervous system. I. Demonstration of monoamines in the cell bodies of brain stem neurons. *Acta Physiol. Scand.* 62(Suppl. 232): 1–55.

Damasio, A. R. (1994). *Descartes' Error: Emotion, Reason and the Human Brain.* New York: Putnam.

Darnal, R. A. (1987). The thermophysiology of the newborn infant. *Med. Instrum.* 21: 16–22.

Dart, R. A. (1964). The ecology of the South African man-apes. In *Ecological Studies in Southern Africa,* D.H. S. Davies (ed.), 49–66. The Hague: H. W. Junk.

Darwin, C. (1871). *The Decent of Man.* London: Murray.

De Vries, A. L., and Y. Lin. (1977). The role of glycoprotein anti-freezes in the survival of Antarctic fishes. In *Adaptations Within Antarctic Ecosystems,* G. A. Llano (ed.), 439–458. Washington, D. C.: Gulf Publishing.

Dean, J. B., and J. A. Boulant. (1989). Effects of synaptic blockade on thermosensitive neurons in rat diencephalon in vitro. *Am. J. Physiol.* 257: R65-R73.

Donald, D. E., D. J. Rowlands, et al. (1970). Similarity of blood flow in the normal and the sympathectomized dog hind limb during graded exercise. *Circ. Res.* 26: 185–199.

Drinkwater, B. L., ed. (1986). *Female Endurance Athletes.* Champaign, IL: Human Kinetics Publishers.

DuBois, E. F. (1949). Why are fever temperatures over 106°F rare? *Am. J. Med. Sci.* 217: 361–368.

Eagan, C. J. (1963). Local vascular adaptation to cold in man. *Fed. Proc.* 22: 947–952.

Eckhardt, R. B. (1987). Was plio-pleistocene hominid brain a pleiotropic effect of adaptation to heat stress? *Anthrop. Anz.* 45: 193–201.

Eichna, L. W., C. R. Park, et al. (1950). Thermal regulation during acclimatization in a hot, dry (desert type) environment. *Amer. J. Physiol.* 163: 585–597.

Engel, P., W. Henze, et al (1984). Psychological and physiological performance during long-lasting work in heat with and without wearing cooling vests. In *Thermal Physiology*, J. R. S. Hales (ed.), 323–326. New York: Raven Press.

Exposito, I., F. Mora, et al. (1995). Dopamine-glutamic acid interaction in the anterior hypothalamus: Modulatory effect of melatonin. *NeuroReport* 6(4): 661–665.

Ezzell, C. (1995). Hot stuff: Medical applications of the heat-shock response. *J. NIH Res.* 7: 42–45.

Feldberg, W., and R. D. Myers. (1964). Effects on temperature of amines injected into the cerebral ventricles. A new concept of temperature regulation. *J. Physiol.* (London) 173: 226–237.

Ferry, D. M., T. J. Butt, et al. (1989). Bacterial chemotactic oligopeptides and the intestinal mucosal barrier. *Gastroenterology* 97: 61–67.

Fialkowski, K. R. (1978). Early hominid brain evolution and stress: A hypothesis. *Stud. Phys. Anthro.* 4: 87–92.

Fialkowski, K. R. (1986). A mechanism for the origin of the human brain: A hypothesis. *Curr. Anthro.* 27: 288–290.

Field, S. B. and J. W. Hand (1990). *An Introduction to the Practical Aspects of Clinical Hyperthermia.* London: Taylor & Francis.

Folk, G.E. (1974) *Textbook of Environmental Physiology.* Philadelphia: Lea & Febiger.

Folk, G. E., W. L. Randall, et al. (1995). Daily rhythms of temperature, heart rate, blood pressure, and activity in the rat. In *Biotelemetry XIII,* C. Cristalli, C. J. Amalaner Jr., and M. R. Neuman (eds.), 246–250. Williamsburg, VA: International Society for Biotelemetry.

Fox, S. W., and K. Dose. (1972). *Molecular Evolution and the Origin of Life.* San Francisco: Freeman.

Froberg, J. E. (1977). Twenty-four-hour patterns in human performance, subjective and physiological variables and differences between morning and evening active subjects. *Biol. Psychol.* 5(2): 119–134.

Fruth, J. M., and C. V. Gisolfi. (1983). Work-heat tolerance in endurance-trained rats. *J. Appl. Physiol.* 54: 249–253.

Gabriel, M., J. Miller, et al. (1977). Unit activity in cingulate cortex and anteroventral thalamus of the rabbit during differential conditioning and reversal. *J. Comp. Physiol. Psychol.* 91: 423–433.

Galen. (1968). *Opera Omnia. Galen's System of Physiology and Medicine,* R. E. Siegel (ed.). New York: Karger.

Garthwaite, J. (1995). Neural nitric oxide signaling. *Trends Neurosci.* 18: 51–52.

Gathiram, P., M. T. Wells, et al. (1987a). Antilipopolysaccharide improves survival in primates subjected to heat stroke. *Circ. Shock* 23: 157–164.

Gathiram, P., M. T. Wells, et al. (1987b). Prevention of endotoxaemia by non-absorbable antibiotics in heat stress. *J. Clin. Path.* 40: 1364–1368.

Gathiram, P., M. T. Wells, et al. (1988). Prophylactic corticosteroid increases survival in experimental heat stroke in primates. *Aviat. Space Environ. Med.* 59: 352–355.

Gehring, W. J., and R. Wehner. (1995). Heat shock protein systhesis and thermotolerance in *Cataglyphis*, an ant from the Sahara desert. *Proc. Natl. Acad. Sci.* 92: 2994–2998.

Gerad, H., D. A. van Echo, et al. (1984). Doxorubicin, cyclophosphamide and whole body hyperthermia for treatment of advanced soft tissue sarcoma. *Cancer* 53: 2585–2591.

Giffin, E. B. (1994). Paleoneurology: Reconstructing the nervous systems of dinosaurs. *DINO FEST.* 229–241.

Ginsberg, M. D., L. L. Sternau, et al. (1992). Therapeutic modulation of brain temperature: Relevance to ischemic brain injury. *Cerebrov. Brain Met. Rev.* 4: 189–225.

Gisolfi, C. V. (1973). Work-heat tolerance derived from interval training. *J. Appl. Physiol.* 35: 349–354.

Gisolfi, C. V., R. D. Matthes, et al. (1991). Splanchnic sympathetic nerve activity and circulating catecholamines in the hyperthermic rat. *J. Appl. Physiol.* 70(4): 1821–1826.

Gisolfi, C.V., and C. B. Wenger. (1984). Temperature regulation during exercise. In *Exercise and Sport Science Reviews*, vol. 12, R. L. Terjung (ed.), 339–372. Lexington, Ma.: D. C. Heath.

Gooden, B. A., and R. Elsner. (1985). What diving animals might tell us about blood flow regulation. *Perspect. Biol. Med.* 28: 465–474.

Graber, C. D., R. B. Reinhold, et al. (1971). Fatal heat stroke. Circulating endotoxins and gram-negative sepsis as complications. *JAMA* 216: 1195–1196.

Granick, S. (1957). Speculations on the origins and evolution of photosynthesis. *Ann. N.Y. Acad. Sci.* 69: 292–308.

Greenberg, N. (1980). Physiological and behavioral thermoregulation in living reptiles. In *A Cold Look at the Warm-Blooded Dinosaurs,* D. K. Thomas and E. C. Olson (eds.), 141–166. Boulder, CO: Westview Press.

Groenink, L. J. C., J. van der Gugten, et al. (1995). Stress-induced hyperthermia in mice: Pharmacological and endocrinological aspects. *Ann. N.Y. Acad. Sci.* 771: 252–256.

Gross, M. (1998). *Life on the Edge: Amazing Creatures Thriving in Extreme Environments.* New York: Plenum Press.

Hales, J. R. S. (1983). Thermoregulatory requirements for regional circulatory adjustments to promote heat loss in animals: A review. *J. Therm. Biol.* 8: 219–224.

Hales, J. R. S., R. W. Hubbard, et al. (1998). Limitation of heat tolerance. *Handbook of Physiology,* Section 4: *Adaptation to the Environment,* M. J. Fregly and C. M. Blatteis (eds.). (American Physiological Society). New York: Oxford University Press.

Hall, D. M., K. R. Baumgardner, et al. (1999). Splanchnic tissues undergo hypoxic stress during whole body hyperthermia. *Amer. J. Physiol.* 276: G1195–G1203.

Hammel, H. T. (1964). *Terrestrial Animals in Cold: Recent Studies of Primitive Man*. Washington, D. C.: American Physiological Society.

Hammel, H. T. (1965). *Neurons and Temperature Regulation*. Philadelphia: Saunders.

Hammel, H. T. (1976). On the origin of endothermy in mammals. *Israel J. Med. Sci.* 12(9): 905–915.

Hammel, H. T., L. I. Crawshaw, et al. (1973). The activation of behavioral responses in the regulation of body temperature in vertebrates. In *The Pharmacology of Thermoregulation,* E. Schonbaum and P. Lomax (eds.), 124–141. Basel: Karger.

Hancock, P. A. (1981). Heat stress impairment of mental performance: A revision of tolerance limits. *Aviat. Space Environ. Med.* 52: 177–180.

Hardy, J. D. (1961). Physiology of temperature regulation. *Physiol. Rev.* 41: 521–606.

Haymaker, W., et al., eds. (1969). *The Hypothalamus*. Springfield: Thomas Books.

Heath, J. E. (1967). Temperature responses of the periodical "17 year" cicada, Magicicada cassini (Homoptera, Cicadidae). *Am. Midland Naturalist* 77(1): 64–76.

Heath, J. E. (1968). The origins of thermoregulation. In *Evolution and Environment,* E. T. Drake (ed.), 259–278. New Haven: Yale University Press.

Heinrich, B. (1974). Thermoregulation in endothermic insects. *Science* 185: 747–756.

Heller, H. C., D. M. Edgar, et al. (1990). Sleep, thermoregulation and circadian rhythms. In *Handbook of Physiology,* Section 4: *Environmental Physiology,* M. J. Fregly and C. M. Blatteis (eds.), 1361–1374 (American Physiological Society). New York: Oxford University Press.

Hellon, R. F. (1975). Monoamines, pyrogens and cations: Their actions on central control of body temperature. *Pharm. Rev.* 26(4): 289–321.

Hellstrom, B., and H. T. Hammel (1967). Some characteristics of temperature regulation in the anesthetized dog. *Am J. Physiol.* 213: 547–556.

Helton, W. S. (1994). The pathophysiologic significance of alterations in intestinal permeability induced by total parenteral nutrition and glutamine. *J. Parent. Enteral. Nut.* 18: 289–290.

Henane, R., A. Buguet, et al. (1977). Variations in evaporation and body temperatures during sleep in man. *J. Appl. Physiol.* 42: 50–55.

Hensel, H. (1981). Thermoreception and temperature regulation. In *Monographs Physiol. Soc.,* 254–308. London: Academic Press.

Hey, E. N., and G. Katz. (1970). The range of thermal insulation in the tissues of newborn baby. *J. Physiol.* (Lond.) 207: 667–681.

Hinckel, P., and K. Schroder-Rosenstock. (1981). Responses of pontine units to skin-temperature changes in the guinea pig. *J. Physiol. (London)* 314: 189–194.

Holmer, I. (1984). Required clothing insulation (IREQ) as an analytical index of cold stress. *Ashrae Trans.* 90: 1116–1128.

Hong, S. K. (1973). Pattern of cold adaptation in women divers of Korea (ama). *Fed. Proc.* 32(5): 1614–1622.

Hong, S. K., C. K. Lee, et al. (1969). Peripheral blood flow and heat flux of Korean women divers. *Fed. Proc.* 28: 1143–1148.

Hong, S. K., and H. Rahn. (1967). The diving women of Korea and Japan. *Sci. Am.* 216(5): 34–43.

Hong, S. K., D. W. Rennie, et al. (1987). Humans can acclimatize to cold: A lesson from Korean women divers. *NIPS* 2: 79–82.

Horne, J. (1988). *Why We Sleep: The Function of Sleep in Human and Other Mammals*. New York: Oxford University Press.

Horowitz, M., A. Maloyan, and J. Shlaier. (1997). HSP 70 kDa dynamics in animals undergoing heat stress superimposed on heat acclimation. *Ann. N.Y. Acad. Sci.* 813: 617–619.

Hubbard, R. W., W. D. Bowers, et al. (1977). Rat model of acute heatstroke mortality. *J. Appl. Physiol.* 42: 809–816.

Hubbard, R. W., W. T. Matthew, et al. (1976). The laboratory rat as a model for hyperthermic syndromes in humans. *Am. J. Physiol.* 231: 1119–1123.

Jaattela, M., and D. Wissing (1992). Emerging role of heat shock proteins in biology and medicine. *Ann. Med.* 24: 249–258.

Jansky, L. (1962). Maximal metabolism and organ thermogenesis in mammals. In *Comparative Physiology of Temperature Regulation*, J. P. Hannon and E. Viereck (eds.), 133–174. Ft. Wainwright, AK: Arctic Aeromedical Laboratory.

Jell, R. M. (1974). Response of rostral hypothalamic neurons to peripheral temperature and animals. *J. Physiol. (London)* 240: 295–307.

Jenning, H. S. (1906). *Behavior of the Lower Organisms*. New York: Columbia University Press.

Jerison, H. J. (1973). *Evolution of the Brain and Intelligence*. London: Academic Press.

Jessen, C. (1998). Brain cooling: An economy mode of temperature regulation in artiodactyls. *NIPS* 13: 281–286.

Jessen, C., and G. Feistkorn. (1984). Some characteristics of core temperature signals in the conscious goat. *Am J. Physiol.* 247: R456–R464.

Jessen, C., and G. Kuhnen. (1992). No evidence for brain stem cooling during face fanning in humans. *J. Appl. Physiol.* 72(2): 664–669.

Johnsen, H. K., A. S. Blix, et al. (1987). Selective cooling of the brain in reindeer. *Am. J. Physiol.* 253: R848–R853.

Johnson, C., and K. J. Edleman. (1992). Malignant Hyperthermia: A Review. *J. Perinatol.* 12: 61–71.

Kandel, E. R., and J. H. Schwartz, (1991). *Principles of Neural Science.* New York: Elsevier.

Kao, T. Y., and M. T. Linn (1996). Brain serotonin depletion attenuates heat-stroke-induced cerebral ischemia and cell death in rats. *J. Appl. Physiol.* 80: 680–684.

Kasting, N. W., W. L. Veale, et al. (1978). Suppression of fever at term of pregnancy. *Nature* 271: 245–246.

Keller, A. D. (1933). Observations on the localization in the brain-stem mechanisms controlling body temperature. *Am. J. Med. Sci.* 185: 746.

Kenney, W. L. (1997). Thermoregulation at rest and during exercise in healthy older adults. *Exercise and Sport Science Reviews,* vol. 25, J. O. Holloszy (ed.), 41–76. Baltimore: Williams and Wilkins.

Kielblock, A. J., N. B. Strydom, et al. (1982). Cardiovascular origins of heatstroke pathophysiology: An anesthetized rat model. *Aviat. Space Environ. Med.* 53: 171–178.

Kilduff, T. S., J. D. Miller, et al. (1990). 14C–2-deoxyglucose uptake in the ground squirrel brain during entrance to and arousal from hibernation. *J. Neurosci.* 10: 2463–2475.

Kleinenberg, N. (1872). *Hydra. Eine Anatomisch-entwicklungsgeschichtliche Untersuchung.* Leigzig: Engelmann.

Kleitman, N. (1963). *Sleep and Wakefulness.* Chicago: University of Chicago Press.

Kluger, M. J. (1979). *Fever: Its Biology, Evolution, and Function.* Princeton, NJ: Princeton University Press.

Kluger, M. J. (1991a). The adaptive value of fever. In *Fever: Basic Mechanisms and Management,* P. Mackowiak (ed.), 105–124. New York: Raven Press.

Kluger, M. J. (1991b). Fever: Role of pyrogens and cryogens. *Physiol. Rev.* 71(1): 93–127.

Knochel, J. P. (1989). Heat stroke and related heat stress disorders. *Disease-a-Month* 35: 301–378.

Knutson, R. M. (1974). Heat production and temperature regulation in eastern skunk cabbage. *Science* 186: 746–747.

Kollias, J., R. Boileau, et al. (1972). Effects of physical conditioning in man on thermal responses to cold air. *Int. J. Biometeor.* 16: 389–402.

Kolval'Zon, V. M., and L. M. Mukhametov (1982). Temperature variations in the brain corresponding to unihemispheric slow wave sleep in dolphins. *J. Evol. Biochem. Physiol.* 18: 307–309.

Krantz, G. S. (1968). Brain size and hunting ability in earliest man. *Curr. Anthro.* 9: 450–451.

Kregel, K. C., and C. V. Gisolfi. (1990). Circulatory responses to vasoconstrictor agents during passive heating in the rat. *J. Appl. Physiol.* 68(3): 1220–1227.

Kregel, K. C., P. T. Wall, et al. (1988). Peripheral vascular responses to hyperthermia in the rat. *J. Appl. Physiol.* 64(6): 2582–2588.

Kreider, M. B. (1961). Effects of sleep deprivation on body temperature. *Fed. Proc.* 20: 214.

Kronenberg, F. (1990). Hot flashes: Epidemiology and physiology. *Ann. N.Y. Acad. Sci.* 592: 52–86.

Kronenberg, F., and J. A. Downey. (1987). Thermoregulatory physiology of menopausal hot flashes: A review. *Can. J. Physiol. Pharmacol.* 65: 1312–1324.

Kuenzel, W. J., and A. Van Tienhoven. (1982). Nomenclature and location of avian hypothalamic nuclei and associated circumventricular organs. *J. Comp. Neurol.* 206: 293–313.

Kuhnen, G., and C. Jessen. (1991). Threshold and slope of selective brain cooling. *Pflugers Arch.* 418: 176–183.

Lee, T. F., F. Mora, et al. (1985). Dopamine and thermoregulation: An evaluation with special reference to dopaminergic pathways. *Neurosci. Behav. Rev.* 9: 589–598.

LeGros Clark, W. E. (1938). *The Hypothalamus.* Edinburgh: Oliver and Boyd.

Leithead, C. S., and A. R. Lind. (1964). *Heat Stress and Heat Disorders.* Philadelphia: Davis.

Lemons, D. E., and L. I. Crawshaw. (1978). Temperature regulation in the Pacific lamprey. *Fed. Proc.* 37: 929.

Lende, R. A. (1964). Representation in the cerebral cortex of a primitive mammal. Sensori-motor, visual, and auditory fields in the Echidna. *J. Neurophysiol.* 27: 37–48.

LeVoyer, T., W. G. Cioffi, et al. (1992). Alterations in intestinal permeability after thermal injury. *Arch. Surg.* 127: 26–30.

Lewis, T. (1941). Observations on some normal and injurious effects of cold upon the skin and underlying tissues. III. Frostbite. *Brit. Med. J.* 2: 869–871.

Li, G. C. (1985). Elevated levels of 70,000 dalton heat shock protein in transiently thermotolerant Chinese hamster fibroblasts and in their stable heat resistant variants. *Int. J. Radiat. Oncol. Biol. Phys.* 11: 165–177.

Lin, M. T. (1997). Heatstroke-induced cerebral ischemia and neuronal damage. *Ann. N.Y. Acad. Sci.* 813: 572–580.

Lipton, J. M., and C. B. Ticknor. (1979). Influence of sex and age on febrile responses to peripheral and central administration of pyrogens in the rabbit. *J. Physiol.* (Lond.) 295: 263–272.

Liu, J. C. (1979). Tonic inhibition of thermoregulation in the decerebrate monkey (Saimiri sciureus). *Exp. Neurol.* 64: 632–648.

Llinas, R., and P. S. Churchland. (1996). *The Mind-Brain Continuum.* Cambridge, MA: MIT Press.

Loewi, O. (1921). Über humorale Übertragbarkeit der Herzenervenwirkung. *Pflugers Arch.* 189: 239–242.

Lomax, P., J. G. Bajorek, et al. (1980). Thermoregulatory changes following the menopause. In *Satellite of 28th International Congress of Physiological Sciences,* A. Szelengi and M. Szekely (eds.). Pecs, Hungary.

Long, D. E. (1979). *The Hajj Today.* Albany: State University of New York Press / Middle East Institute.

Lundberg, J. M., and T. Hokfelt. (1983). Coexistence of peptides and classical neurotransmitters. *Trends Neurosci.* 6:325–333.

Lyman, C. P. (1958). Oxygen consumption, body temperature, and heart rate of woodchucks entering hibernation. *Am. J. Physiol.* 194: 83–91.

Mackowiak, P. A. (1994). Fever: Blessing or Curse: A Unifying Hypothesis. *Ann. Int. Med.* 120(12): 1037–1040.

Mackowiak, P. A., and J. A. Boulant. (1996). Fever's Glass Ceiling. *Clin. Infect. Dis.* 22: 525–536.

Madara, J. L., and J. Stafford. (1989). Interferon-gamma directly affects barrier function of cultured intestinal epithelial monolayers. *J. Clin. Invest.* 83: 724–727.

Majno, G. (1975). *The Healing Hand: Man and Wound in the Ancient World.* Cambridge, MA: Harvard University Press.

Malamud, N., W. Haymaker, et al. (1946). Heatstroke: A clinicopathologic study of 125 fatal cases. *Mil. Surg.* 99: 397–449.

Malan, A. (1966a). Suppression de l'hibernation et modification de la thermoregulation par lésions hypothalamiques chez le hamster d'Europe (*Cricetus cricetus*). *Compt. Rendus Soc. Biol.* 160: 1497–1499.

Malan, A. (1966b). Suppression de l'hibernation par lésions hypothalamiques postérieures chez le hamster d'Europe (*Cricetus cricetus*). *J. Physiol.* (Paris) 58: 565–566.

Maron, M. B., J. A. Wagner, et al. (1977). Thermoregulatory responses during competitive marathon running. *J. Appl. Physiol.* 42: 909–914.

McConaghy, F. F., J. R. S. Hales, et al. (1995). Selective brain cooling in the horse during exercise and environmental heat stress. *J. Appl. Physiol.* 79: 1849–1854.

McGinty, D., and R. Szymusiak. (1990). Keeping cool: A hypothesis about the mechanisms and functions of slow-wave sleep. *TINS* 13(12): 480–487.

McGowan, C. (1991). *Dinosaurs, Spitfires and Sea Dragons.* Cambridge, MA: Harvard University Press.

McLean, P. D. (1952). Some psychiatric implications of physiological studies on front temporal portion of limbic system (visceral brain). *Electroenceph. Clin. Neurphysiol.* 4: 407–418.

Mihailovic, L. T. (1972). Cortical and subcortical electrical activity in hibernation and hypothermia. In *Hibernation and Hypothermia: Perspectives and Challenges,* F. E. South, J. P. Hannon et al. (eds.), 487–535. Amsterdam: Elsevier.

Mitchell, H. K., G. Moller, et al. (1979). Specific protection from phenocopy induction by heat shock. *Dev. Genet.* 1: 181–192.

Molnar, G. W., A. L. Hughes, et al. (1973). Effect of skin wetting on finger cooling and freezing. *J. Appl. Physiol.* 35: 205–207.

Moltz, H. (1993). Fever: Causes and consequences. *Neurosci. and Biobehav. Rev.* 17: 237–269.

Mora, F. (1977). *The Neural Basis of Feeding, Drinking and Reward.* Doctoral thesis, Oxford University.

Mora, F. (1995). *El problema Cerebromente.* Madrid: Alianza Editorial.

Mora, F. (ed.) (1996). *El Cerebro Intimo.* Ariel: Barcelona.

Mora, F. (Ed.). (1999). *La Razon Emocional del Cerebro.* Madrid: Arbor.

Mora, F. (1999). The brain and the mind. In *The New Oxford Textbook of Psychiatry,* M. G. Gelder, J. J. Lopez-Ibor, and N. Andreasen (eds.). Oxford: Oxford University Press.

Mora, F., and M. Cobo. (1990). The neurobiological basis of prefrontal cortex self-stimulation: A review and an integrative hypothesis. *Prog. Brain Res.* 85: 419–431.

Mora, F., and A. M. Sanguinetti. (1994). *Diccionario de Neurociencias.* Madrid: Alianza Editorial.

Morrison, P. R., and F. A. Ryser. (1952). Weight and body temperatures in mammals. *Science* 116: 231–232.

Moseley, P. L. (1994). Mechanisms of heat adaptation: Thermotolerance and acclimatization. *J. Lab. Clin. Med.* 123: 48–52.

Moseley, P. L., C. Gapen, et al. (1994). Thermal stress induces epithelial permeability. *Am. J. Physiol.* 267: C425–C434.

Muller, A. W. (1995). Were the first organisms heat engines? A new model for biogenesis and the early evolution of biological energy conversion. *Prog. Biophys. Molec. Biol.* 63: 193–231.

Myers, R. D. (1980). Hypothalamic control of thermoregulation. Neurochemical mechanisms. In *Handbook of the Hypothalamus,* P. J. Morgane and J. Panksepp (eds.), 83–210. New York: Marcel Dekker.

Myers, R. D., and T. F. Lee. (1989). Neurochemical aspects of thermoregulation. In *Advances in Comparative and Environmental Physiology,* L. C. H. Wang (ed.), 162–194. Berlin: Springer-Verlag.

Myers, R. D., and W. L. Veale. (1970). Body temperature: Possible ionic mechanism in the hypothalamus controlling the set point. *Science* 170: 95–97:

Myhre, K., and H. T. Hammel. (1969). Behavioral regulation of internal temperature in the lizard, Tiliqua scincoides. *Am. J. Physiol.* 217: 1490–1495.

Nadel, E. R. (1987). Comments on "Keeping a cool head." *NIPS* 2: 33–34.

Nauts, H. C. (1982a). *Bacterial Products in the Treatment of Cancer: Past, Present and Future.* London and New York: Academic Press.

Nauts, H. C. (1982b). *Bacterial Pyrogens: Beneficial Effects on Cancer Patients.* New York: Alan R. Liss.

Nelson, D. O., and C. L. Prosser. (1979). Effect of preoptic lesions on behavioral thermoregulation of green sunfish, Lepomis cyanellus, and of goldfish, Carassius auratus. *J. Comp. Physiol.* 129: 193–197.

Niazi, S. A., and F. J. Lewis. (1958). Profound hypothermia in man, report of a case. *Am. Surg.* 147: 264–266.

Nielsen, B. (1988). Natural cooling of the brain during outdoor bicycling? *Pflugers Arch.* 411: 456–461.

Nielsen, B., G. Savard, et al. (1990). Muscle blood flow and muscle metabolism during exercise and heat stress. *J. Appl. Physiol.* 69(3): 1040–1046.

Nielsen, B., J. R. S. Hales, et al. (1993). Human circulatory and thermoregulatory adaptations with heat acclimation and exercise in a hot, dry environment. *J. Physiol. (London)* 460: 467–485.

Obal, F. J. (1984). Thermoregulation and sleep. In *Sleep Mechanisms,* A. Borberly and J. L. Valaty (eds.). *Exp. Brain Res. Suppl.* 8: 157–172.

O'Dwyer, S. T., H. R. Michie, et al. (1988). A single dose of endotoxin increases intestinal permeability in healthy humans. *Arch. Surg.* 123: 1459–1464.

Olds, J., and P. Milner (1954). Positive reinforcement produced by electrical stimulation of septal area and other regions of rat brain. *J. Comp. Physiol. Psychol.* 47: 419–427.

Olsson, Y., H. S. Sharma, et al. (1995). The opioid receptor antagonist naloxone influences the pathophysiology of spinal cord injury. In *Neuropeptides in the Spinal Cord,* F. Nyberg, H. S. Sharma, and Z. Wiesenfeld (eds.), 381–399. Amsterdam: Elsevier.

Ostrom, J. H. (1980). The evidence for endothermy in dinosaurs. In *A Cold Look at the Warm-Blooded Dinosaurs,* D. K. Roger, T. Olsen, and E. C. Olson (eds.), 15–54. Boulder, CO: West View Press.

Paladino, F. V., and J. R. Spotila. (1994). Does the physiology of large living reptiles provide insights into the evolution of endothermy and paleophysiology of extinct dinosaurs? *DINO FEST.* 261–273.

Pantin, C. F. A. (1956). The origin of the nervous system. *Pub. Staz. Zool. Napoli* 28: 171–181.

Papez, J. W. A. (1937). A proposed mechanism of emotion. *Arch. Neurol. Psychiat.* 38: 725–743.

Parker, G. H. (1919). *The Elementary Nervous System.* Philadelphia: Lippincott.

Parmegiani, P. L. (1980). Temperature regulation during sleep: A study in homeostasis. In *Physiology in Sleep,* J. Onens and C. D. Barnes (eds.), 97–143. New York: Academic Press.

Parmegiani, P. L., and C. Rabini. (1967). Shivering and panting during sleep. *Brain Res.* 6: 789–791.

Parmegiani, P. L., and C. Rabini (1970). Sleep and environmental temperature. *Arch. Ital. Biol.* 108: 369–387.

Passano, L. M. (1963). Primitive nervous systems. *Proc. Natl. Acad. Sci.* 50: 306–313.

Pierau, R. K., J. Schenda, et al. (1994). Possible implications of the plasticity of temperature-sensitive neurons in the hypothalamus. In *Thermal Balance in Health and Disease,* E. Zeisberger, E. Schonbaum, and P. Lomax (eds.), 31–37. Basel: Birkhauser Verlag.

Prosser, C. L. 1961. Temperature. In *Comparative Animal Physiology*, 2nd ed., C. L. Prosser and F. A. Brown, Jr. (eds.), 238–284. Philadelphia: W. B. Saunders.

Pugh, L. G. C. E., J. L. Corbett, et al. (1967). Rectal temperatures, weight losses, and sweat rates in marathon running. *J. Appl. Physiol.* 23(3): 347–352.

Quan, N., L. Xin, et al. (1992). Preoptic norepinephrine-induced hypothermia is mediated by a2-adrenoceptors. *Am. J. Physiol.* 262: R407–R411.

Ranson, S. W. J. (1940). Regulation of body temperature. *Pub. Inst. Neurol. Chicago: Northwestern Univ. Med. Sch.* 12: 394.

Rao, P. D. P., N. Subhedar, et al. (1981). Cytoarchitectonic pattern of the hypothalamus in the cobra, Naja naja. *Cell Tissue Res.* 217: 505–529.

Rautenberg, W., R. Neacker, and B. May (1972). Thermoregulatory responses of the pigeon to changes of the brain and spinal cord temperatures. *Pflugers Arch.* 338: 31–42.

Renbourn, E. T. (1960). Body temperature and pulse rate in boys and young men prior to sporting contests: With a review of the literature. *J. Psychosom. Res.* 4: 149–175.

Reynolds, W. W., and M. E. Casterlin. (1976). Behavioural fever in teleost fishes. *Nature* 259: 41–42.

Riabowol, K. T., L. A. Mizzen, et al. (1988). Heat shock is lethal to fibroblasts microinjected with antibodies against hsp70. *Science* 242: 433–436.

Rietschel, E. T., and H. Brade. (1992). Bacterial endotoxins. *Sci. Am.* (Aug.): 54–61.

Roberts, M. F., C. B. Wenger, et al. (1977). Skin blood flow and sweating changes following exercise training and heat acclimation. *J. Appl. Physiol.* 43: 133–137.

Robinson, S. (1963). Temperature regulation in exercise. *Pediatrics* 32(4): 691–702.

Rolls, E. T., J. M. Burton, et al. (1976). Hypothalamic neuronal responses associated with the sight of food. *Brain Res.* 111: 53–66.

Rolls, E. T., J. M. Burton, et al. (1980). Neurophysiological analysis of brain stimulation reward in the monkey. *Brain Res.* 194: 339–357.

Romer, A. S. (1937). The braincase of the Carboniferous crossopterygian Megalichthys nitidus. *Bull. Mus. Comp. Zool.* 82: 1–73.

Romer, A. S. (1970). *The Vertebrate Body,* Philadelphia: W. B. Saunders.

Rothman, S. M., and J. W. Olney. (1986). Glutamate and the pathophysiology of hypoxic-ischemic brain damage. *Ann. Neurol.* 19: 105–111.

Rowell, L. B. (1986). *Human Circulation: Regulation During Physical Stress.* New York: Oxford University Press.

Ruben, J. (1995). The evolution of endothermy in mammals and birds: From physiology to fossils. *Ann. Rev. Physiol.* 57: 69–95.

Rusack, B. (1989). The mammalian circadian system: Models and physiology. *J. Biol. Rhythms* 4: 121–134.

Ryan, A. J., S. W. Flanagan, et al. (1992). Acute heat stress protects rats against endotoxin shock. *J. Appl. Physiol.* 73(4): 1517–1522.

Ryberg, W. (1987). A year after dying, Iowa tot is healthy, happy, full of life. *Des Moines Sunday Register,* 22 Feb., 1A, 7A.

Sakurada, S., and J. R. S. Hales (1998). A role for gastrointestinal endotoxins in the enhancement of heat tolerance by phsical fitness. *J. Appl. Physiol.* 84: 207–214.

Saltin, B., and L. Hermansen. (1966). Esophageal, rectal, and muscle temperature during exercise. *J. Appl. Physiol.* 21:1757–1762.

Satinoff, E. (1964). Behavioral thermoregulation in response to local cooling of the rat brain. *Am. J. Physiol.* 206: 1389–1394.

Satinoff, E. (1967). Disruption of hibernation caused by hypothalamic lesions. *Science* 155: 1031–1033.

Satinoff, E. (1978). Neural organization and evolution of thermal regulation in mammals. *Science* 201: 16–22:

Satinoff, E., and R. A. Prosser. (1988). Suprachiasmatic nuclear lesions eliminate circadian rhythms of drinking and activity but not of body temperature in rats. *J. Bio. Rhythms* 3: 1–22.

Satinoff, E., and J. Rutstein. (1971). Behavioral thermoregulation in rats with anterior hypothalamic lesions. *J. Comp. Physiol. Psychol.* 77: 302–312.

Sato, F., M. Owen, et al. (1990). Functional and morphological changes in the eccrine sweat gland with heat acclimation. *J. Appl. Physiol.* 69(1): 232–236.

Sato, H. (1984). Midbrain neurons of rats responsive to hypothalamic temperature and their local thermosensitivity. *J. Therm. Biol.* 9: 39–45.

Sato, K., and F. Sato. (1983). Individual variations in structure and function of human eccrine sweat gland. *Am. J. Physiol.* 245: R203–R208.

Sato, K. T., N. L. Kane, et al. (1996). Reexamination of tympanic membrane temperature as a core temperature. *J. Appl. Physiol.* 80: 1233–1239.

Schmidt, I. (1978). Behavioral and autonomic thermoregulation in heat stressed pigeons modified by central thermal stimulation. *J. Comp. Physiol.* 127: 75–87.

Scholander, P. F., V. Walters, R. Hock, and L. Irving. (1950). Body insulation of some arctic and tropical mammals and birds. *Biol. Bull.* 99: 225–236.

Scott, I. M., and J. A. Boulant. (1984). Dopamine effects on thermosensitive neurons in hypothalamic tissue slices. *Brain Res.* 306: 157–163.

Segovia, G., A. D. Arco, et al. (1997). Endogenous glutamate increases extracellular concentrations of dopamine, GABA, and taurine through NMDA and AMPA/kainate receptors in striatum of the freely moving rat: A microdialysis study. *J. Neurochem.* 69: 1476–1483.

Shapiro, Y., A. Magazanik, et al. (1979). Heat intolerance in former heatstroke patients. *Ann. Int. Med.* 90: 913–916.

Sharma, H. S., J. Westman, et al. (1997). Opioid receptor antagonists attenuate heat stress-induced reduction in cerebral blood flow, increased blood-brain-barrier permeability, vasogenic edema and cell changes in the rat. In *Thermoregulation: Tenth International Symposium on the Pharmacology of Thermoregulation,* C. M. Blatteis (ed.), 559–571. New York: New York Academy of Sciences.

Sheehan, P. M. (1994). The extinction of dinosaurs. *DINO FEST.* 411–423.

Shibolet, S. (1962). The Clinical Picture of Heatstroke. In *Proceedings of the Tel Hashomer Hospital,* 80–93. Tel-Aviv.

Shibolet, S., M. C. Lancaster, et al. (1976). Heat stroke: A review. *Aviat. Space Environ. Med.* 47: 280–301.

Shochina, M. W. H., U. Meiri, and M. Horowitz. (1996). Heat acclimation and hypohydration in aged rats: The involvement of adrenergic pathways in thermal-induced vasomotor responses in the portal circulation. *J. Therm. Biol.* 21: 289–295.

Simon, E., F. K. Pierau, and D. C. M. Taylor. (1986). Central and peripheral thermal control of effectors in homeothermic temperature regulation. *Physiol. Rev.* 66: 235–300.

Simon, H. B. (1994). Hyperthermia and heatstroke. *Hosp. Prac.* 29: 65–80.

Slonim, N. B., ed. (1974). *Environmental Physiology.* St. Louis: C. V. Mosby.

Snell, E. S., and E. Atkins. (1968). The mechanisms of fever. In *The Biological Basis of Medicine,* E. E. Bittar, and N. Bittar (eds.), 397–419. New York: Academic Press.

Sohnle, P. G., and S. R. Gambert. (1982). Thermoneutrality: Evolutionary advantage against ageing? *Lancet* May 15: 1099–1100.

Somasundaram, S., H. Hayllar, et al. (1995). The biochemical basis of non-steroidal anti-inflammatory drug-induced damage to the gastrointestinal tract: A review and a hypothesis. *Scand. J. Gastroenterol.* 30: 289–299.

Spotila, J. R. (1980). Constraints of body size and environment on the temperature regulation of dinosaurs. In *A Cold Look at the Warm-Blooded Dinosaurs,* R. D. K. Thomas and E. C. Olson (eds.), 223–252. Boulder, CO: Westview Press.

Stevens, E. D. (1973). The evolution of endothermy. *J. Theor. Biol.* 38: 597–611.

Stitt, J. T. (1979). Fever versus hyperthermia. *Fed. Proc.* 38: 39–43.

Streffer, C. (1995). *Molecular and Cellular Mechanisms of Hyperthermia.* New York: Springer-Verlag.

Strumwasser, F. (1959). Regulatory mechanisms, brain activity, and behavior during deep hibernation in the squirrel, Citellus beecheyi. *Am. J. Physiol.* 196: 23–30.

Strydom, N. B. (1980). Heat intolerance: Its detection and elimination in the mining industry. *S. Afr. J. Sci* 76: 154–156.

Sweatman, P., and R. M. Jell. (1977). Dopamine and histamine sensitivity of rostral hypothalamic neurons in the cat: Possible involvement in thermoregulation. *Brain Res.* 127: 173–178.

Tashkim, D. P., P. J. Goldstein, and D. H. Simmons. (1972). Hepatic lactate uptake during decreased liver perfusion and by hyposemia. *Am. J. Physiol.* 223: 968–974.

Taylor, C. R. (1966). The vascularity and possible thermoregulatory function of the horns in goats. *Physiol. Zool.* 39: 127–139.

Taylor, C. R. (1970). Dehydration and heat: Effects on temperature regulation of East African ungulates. *Am. J. Physiol.* 219: 1136–1139.

Taylor, C. R., and C. P. Lyman. (1972). Heat storage in running antelopes: Independence of brain and body temperatures. *Am. J. Physiol.* 222: 114–117.

Thomas, R. D. K., and E. C. Olson. (1980). *A Cold Look at the Warm-Blooded Dionsaurs.* Boulder, CO: Westview Press.

Thompson, R. H. (1959). Influence of environmental temperature upon pyrogenic fever. PhD thesis, University of Pennsylvania.

Tissieres, A., H. K. Mitchell, et al. (1974). Protein synthesis in salivary glands of *Drosophila melanogaster*: Relation to chromosome puffs. *J. Mol. Biol.* 84: 389–398.

Tobias, P. V. (1971). *The Brain in Hominid Evolution.* New York: Columbia University Press.

Tolaas, J. (1978). REM sleep and the concept of vigilance. *Biol. Psychiatry* 13: 135–148.

Ulmasov, K. A., S. Shammakov, et al. (1992). Heat shock proteins and thermoresistance in lizards. *Proc. Natl. Acad. Sci* 89: 1666–1670.

van der Zee, J., G. C. van Rhoon, et al. (1990). Clinical hyperthermic practice: Whole-body hyperthermia. In *An Introduction to the Practical Aspects of Clinical Hyperthermia,* S. B. Field and J. W. Hand (eds.), 185–212. London: Taylor & Francis.

Van Holde, K. E. (1980). *The Origins of Life and Evolution.* New York: Alan R. Liss.

Van Valen, K. (1960). Therapsids as mammals. *Evolution* 14: 304–313.

von Neumann, J. (1963). Probabilistic logic and the synthesis of reliable organisms for unreliable components. In his *Collected Works,* vol. 5, 329–378. London: Pergamon Press.

Walker, J. M., S. F. Glotzbach, et al. (1977). Sleep and hibernation in ground squirrels (Citellus spp.): Electrophysiological observations. *Am. J. Physiol.* 223: R213–R221.

Wang, L. C., Y. Cui, et al. (1994). Is the hypothalamic serotonergic system involved in septal 5-HT induced hypothermia in rats? In *Thermal Balance in Health and Disease: Advances in Pharmacological Sciences,* E. Zeisberger, E. Schonbaum, and P. Lomax (eds.), 469–473. Basel: Birkhauser Verlag.

Watanabe, T., A. Morimoto, et al. (1986). Effect of amine on temperature-responsive neuron in slice preparation of rat brain stem. *Am. J. Physiol.* 250: R553–R559.

Weiss, B., and V. G. Laties (1961). Behavioral thermoregulation. *Science* 133: 1338–1344.

Wenger, C. B. (1987). More comments on "Keeping a cool head." *NIPS* 2: 150–151.

Wenger, C. B., M. F. Roberts, et al. (1976). Nocturnal lowering of thresholds for sweating and vasodilation. *J. Appl. Physiol.* 41: 15–19.

Weshler, Z., D. S. Kapp, et al. (1984). Development and decay of systemic thermotolerance in rats. *Cancer Res.* 44: 1347–1351.

Whitby , J. D., and L. J. Dunkin. (1972). Cerebral, oesophageal, and nasopharyngeal temperatures. *Br. J. Anaesth.* 43: 673–676.

Withers, P. C. (1992). *Comparative Animal Physiology.* Fort Worth, TX: Saunders College Press.

Yost, R. M. (1950). Sydenham's philosophy of science. *Osiris* 9: 84–105.

Young, M. S., and D. Kindred. (1993). Malignant hyperthermia: Not just an operating room emergency. *Medsurg. Nursing* 2: 41–46.

Zeisberger, E. (1987). The role of monoaminergic neurotransmitters in thermoregulation. *Can. J. Physiol. Pharmacol.* 65: 1395–1401.

Zeisberger, E., and J. Roth. (1996). Central regulation of adaptive responses to heat and cold. In *Handbook of Physiology,* M. J. Fregley and C. M. Batteis (eds.), 579–593 (American Physilogical Society). New York: Oxford University Press.

Zethof, T. J. J., J. van der Heyden, et al. (1994). Stress-induced hyperthermia in mice: A methodological study. *Physiol. Behav.* 55: 109–115.

Ziegler, T. R., R. J. Smith, et al. (1988). Increased intestinal permeability associated with infection in burn patients. *Arch. Surg.* 123: 1313–1319.

Index